Improving Joint Expeditionary Medical Planning Tools Based on a Patient Flow Approach

Edward W. Chan, Heather Krull, Beth E. Lachman,
Tom LaTourrette, Rachel Costello, Don Snyder,
Mahyar A. Amouzegar, Hans V. Ritschard,
D. Scott Guermonprez

Prepared for the United States Air Force

Approved for public release; distribution unlimited

PROJECT AIR FORCE

The research described in this report was sponsored by the United States Air Force under Contract FA7014-06-C-0001. Further information may be obtained from the Strategic Planning Division, Directorate of Plans, Hq USAF.

Library of Congress Cataloging-in-Publication Data

Improving joint expeditionary medical planning tools based on a patient flow approach /
Edward W. Chan ... [et al.].
   p. cm.
  Includes bibliographical references.
  ISBN 978-0-8330-5900-0 (pbk. : alk. paper)
1. Battle casualties—Medical care—Planning. 2. Battle casualties—Medical care—Mathematical models. 3. United States—Armed Forces—Medical care—Planning. 4. United States—Armed Forces—Medical care— Mathematical models. 5. Medicine, Military—United States—Planning. I. Chan, Edward Wei-Min, 1970-

  UH394.I67 2912
  355.3'450684—dc23

                                                                           2012000928

The RAND Corporation is a nonprofit institution that helps improve policy and decisionmaking through research and analysis. RAND's publications do not necessarily reflect the opinions of its research clients and sponsors.

RAND® is a registered trademark.

Published 2012 by the RAND Corporation
1776 Main Street, P.O. Box 2138, Santa Monica, CA 90407-2138
1200 South Hayes Street, Arlington, VA 22202-5050
4570 Fifth Avenue, Suite 600, Pittsburgh, PA 15213-2665
RAND URL: http://www.rand.org/
To order RAND documents or to obtain additional information, contact
Distribution Services: Telephone: (310) 451-7002;
Fax: (310) 451-6915; Email: order@rand.org

Preface

The Air Force Medical Service supports combatant commanders by providing treatment at forward-deployed medical treatment facilities and aeromedical evacuation to patients who need higher levels of care. This concept of operations involves a rapid flow of patients across different treatment facilities, which requires close integration of treatment and evacuation functions. Balancing the deployment of treatment and evacuation resources is therefore necessary to ensure that the right mix of resources is available in a timely fashion.

The Air Force Surgeon General asked the RAND Corporation to identify methods to improve the planning of forward-deployed medical resources, taking into account the variety of scenarios that the U.S. military may face in future operations. To do so, RAND developed an approach that integrates planning for medical resources across treatment and evacuation functions, across levels of care, and across the military services. The approach is built around a RAND-developed construct, the stabilization, triage and treatment, and evacuation of patients (STEP) rate. Essentially a patient flow rate, the STEP rate can help planners better understand the interdependencies among treatment and evacuation functions, levels of care, and the military services. This understanding will help planners ensure that the right level of resources is sent to the right areas, thereby preventing resource imbalances that impede the rapid movement of patients across the entire medical system. The STEP rate concept is compatible with existing joint medical planning tools. Recommended modifications to these tools are discussed in this report.

Related RAND research is presented in the following publications:

- *How Should Air Force Expeditionary Medical Capabilities Be Expressed?* Don Snyder, Edward W. Chan, James J. Burks, Mahyar A. Amouzegar, and Adam C. Resnick (MG-785-AF). This report presents a new metric for measuring expeditionary medical support and develops a framework for applying it across three Air Force medical mission areas: deployed support to the warfighter, humanitarian relief, and defense support to civil authorities.
- *New Equipping Strategies for Combat Support Hospitals*, Matthew W. Lewis, Aimee Bower, Mishaw T. Cuyler, Rick Eden, Ronald E. Harper, Kristy Gonzalez Morganti, Adam C. Resnick, Elizabeth D. Steiner, and Rupa S. Valdez (MG-887-A). This report describes a new equipping strategy for the Army's Combat Support Hospitals.

The research presented in this report was conducted as part of a fiscal year 2009 study titled "Improved Planning for Wartime Requirements for the Air Force Medical Service." The work was sponsored by the Air Force Surgeon General and conducted within the Resource

Management Program of RAND Project AIR FORCE. This report should be of interest to medical planners and programmers in the Air Force and in the joint community.

RAND Project AIR FORCE

RAND Project AIR FORCE (PAF), a division of the RAND Corporation, is the U.S. Air Force's federally funded research and development center for studies and analyses. PAF provides the Air Force with independent analyses of policy alternatives affecting the development, employment, combat readiness, and support of current and future air, space, and cyber forces. Research is conducted in four programs: Force Modernization and Employment; Manpower, Personnel, and Training; Resource Management; and Strategy and Doctrine.

Additional information about PAF is available on our website:
http://www.rand.org/paf/

Contents

Figures

Tables

Summary

The current concept of operations (CONOPS) for expeditionary medical care emphasizes quickly moving patients to a series of successively more sophisticated medical facilities that provide the patients with the care necessary to ultimately treat their injuries or conditions. This process requires close coordination between the treatment facilities and the evacuation resources that link them. However, the processes and tools currently used in planning for expeditionary medical resources do not fully reflect the current CONOPS. Currently, the capability of treatment facilities is typically measured and expressed in terms of the number of hospital beds in the facility, and aeromedical evacuation capabilities are typically measured and expressed in terms of the number of teams or aircraft available. Thus, planning is not aligned with operational practice and is not well integrated across the full spectrum of echelons and functions.

Expressing treatment and evacuation capabilities in terms of such measures as the numbers of beds and aircraft has two disadvantages. The first is that such measures are static measures of *capacity*: Beds and aircraft are fundamentally measures of the numbers of items. However, what is of concern to planners is not the number of items at each facility or function but rather the *capability* that can be provided by those resources. The second disadvantage is that the treatment and evacuation functions use different units of measure. With treatment resources being measured in beds and evacuation assets being measured in aircraft, it is not readily apparent how many aircraft are necessary to provide support to a field hospital of a given size.

We propose a planning concept that is consistent with the military medical CONOPS and that helps integrate medical planning across treatment and evacuation functions, across the increasing levels of care, and across the different military services. Our concept begins by proposing that treatment and evacuation functions at all levels use patient flow rate as the common unit of measurement. The goal of every treatment facility and every evacuation asset is the stabilization, triage and treatment, and evacuation of patients (STEP) to the next and higher level of care as quickly as is prudent. Therefore, a measure of the capability of a component to provide care is the rate at which that component can carry out these activities.

We propose that the patient flow rate, or STEP rate,[1] be applied across the entire medical network. For treatment facilities, the STEP rate would be the number of patients treated per

---

[1]   RAND introduced the STEP rate in an earlier report (D. Snyder, E. W. Chan, J. J. Burks, M. A. Amouzegar, and A. C. Resnick, *How Should Air Force Expeditionary Medical Capabilities Be Expressed?* Santa Monica, Calif.: RAND Corporation, MG-785-AF, 2009) that argues why the STEP rate is a more appropriate unit of measurement than beds and illustrates how the STEP rate would apply in different scenarios and configurations of the medical network. The authors suggest that treatment facility unit type codes be re-expressed in terms of STEP rates.

unit of time; for evacuation assets, the STEP rate would be the number of patients evacuated per unit of time. This approach has the following advantages:

- It corresponds to the CONOPS of flow of patients away from the point of injury and toward facilities that can provide proper care.
- It applies across all resources, levels of care, and the military services.
- It measures capability rather than capacity. That is, rather than measuring resource levels (e.g., the number of beds, aircraft, and medical staff), the patient flow rate measures the capability provided by the resources.

The CONOPS of patient care can be likened to the flow of fluid through a system of valves and holding tanks. In such a system, the valves govern the rate at which fluid flows into and out of the holding tanks. Ideally, the flow of fluid out of a tank is at least as fast as the flow of fluid into a tank; otherwise, fluid backs up in the tank. Figure S.1 illustrates this analogy with an example of a medical deployment laydown.

In the figure, the flow starts with patients entering the medical facilities of the Air Force, Army, or Navy and Marine Corps because they have been wounded in action (WIA) as a result of combat or enemy action or because they are suffering from a form of disease or nonbattle injury (DNBI). Different scenarios will produce casualties at different rates for the different services and yield different mixes of injuries and illnesses. The mix of injuries and illnesses can have a major effect on the patient flow rate that can be achieved with a given set of resources.

The flow of patients next progresses to the first treatment facility, which is represented by two holding tanks. The contents of the first tank represent patients who arrive requiring stabilization or treatment; this tank can be thought of as the waiting room of a hospital emergency department. The first valve represents the rate at which patients receive care at the facility, and it governs the rate at which the first tank is emptied. The rate at which patients receive care is a function of the treatment resources available, including capacity at the emergency room, operating room, and intensive care unit and the capacity of the staff. Patients who have been

Figure S.1
Analogy of Patient Flow in a System of Holding Tanks and Valves

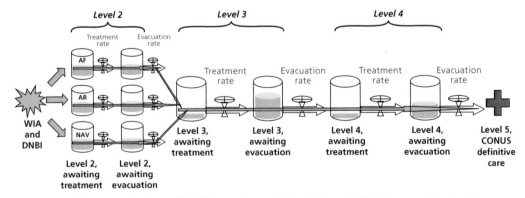

NOTE: AF = Air Force. AR = Army. CONUS = continental United States. NAV = Navy and Marine Corps.
RAND *TR1003-S.1*

treated and are stable enough for transport are moved to the second tank, where they await evacuation to the next-higher level of care. This second tank corresponds to holding beds at the treatment facility or at an aeromedical staging facility.[2] The rate at which patients leave this holding tank is governed by the rate at which the aeromedical evacuation system (or some other transport resource) can transport them.[3] From there, the patient flow continues on to higher levels of care.

Actual operations may be more complex, involving multiple locations and multiple paths of flow; patients may skip levels of care or remain at one level until they are released. Even in these situations, the principles of flow rate apply. Regardless of the specific path, the system still requires coordination between treatment and evacuation resources to ensure that the flow of patients is not unnecessarily delayed.

For the planner seeking to put together into a medical system a series of treatment facilities linked by evacuation assets, the goal is to prevent resource imbalances that impede the rapid movement of patients across the entire medical system. The STEP rate helps by allowing planners to better understand the interdependencies among treatment and evacuation functions, levels of care, and the military services. With this understanding, planners will be better able to assign resources in a way that ensures that these elements work together.

The STEP rate shows how shortages in resources in one area affect resources needed in other areas. For example, reductions in the number of available airlifters would reduce the evacuation rate between treatment facilities to a number below the rate needed to keep up with the flow of casualties. This reduced evacuation rate would mean that patients would have to wait longer at the lower-level facility. Consequently, planners would need to increase holding capacity at that location in the form of holding beds within the theater hospital facility.

The STEP rate concept also illustrates the importance of coordination between the services. The Air Force, for example, needs to know the casualty stream from Army and Marine Corps units to plan for the right number of aeromedical evacuation flights, crews, and patient movement items. Conversely, it is also vital that Army and Navy medical planners be aware of Air Force capabilities for aeromedical evacuation because these capabilities could affect the holding capacity required at Army and Navy forward field hospitals and casualty receiving ships.

As is the case with using any method for estimating medical resource requirements, using the patient flow rate or STEP rate concept requires an understanding of the number, type, and timing of casualties in the scenario. If the estimates of casualties are inaccurate, the resulting estimates of medical resources required will be inaccurate, regardless of the resource estimation algorithm. Although no casualty forecast is likely to predict with certainty the events that will unfold in an actual contingency, efforts can and should be made to develop forecasts that are appropriate to the different types of scenarios the U.S. military may face in the future.

The primary medical planning tool approved by the Joint Staff for use by combatant commands for developing their operational plans is the Joint Medical Analysis Tool (JMAT),

---

[2]   Patients in the holding facility are likely to require continuing medical care and therefore to consume medical resources. The distinction we make is that some patients in the holding facility are undergoing treatment because they are not yet well enough to be transported, whereas other patients in the holding facility are there only because they are waiting for transportation to become available. The latter patients are the ones who would be in the "awaiting evacuation" tank.

[3]   Transportation is not continuous but rather usually occurs in discrete vehicle loads. Therefore, a more exact analogy might be a valve that periodically opens, releasing a measured amount of fluid each time.

which is managed by the Joint Staff Theater Medical Information Program. (As of September 2010, the version of JMAT currently in use is 1.0.1.0.) Because JMAT simulates the flow of patients within and across facilities, it already operates in a manner consistent with the patient flow rate concept just described, even though it does not explicitly express treatment or evacuation capabilities in terms of the STEP rate. To more fully implement the STEP concept, three main modifications to JMAT would be required:

- **Extend JMAT to include the entire chain of care.** The flow concept we present in this report emphasizes the interdependence among treatment and evacuation functions and among treatment facilities in different echelons. This interdependence requires a planning tool that encompasses all levels of care. JMAT 1.0.1.0 models Levels 3, 4, and 5. We recommend that JMAT be extended to include support for modeling Level 2 care. In addition, we recommend that JMAT allow more flexibility in the paths taken by patients, including the fact that they may skip echelons and that stopover locations in between echelons may be involved. (The latter may be necessary in scenarios in which the flying time between the theater and definitive care facilities is long.)
- **Enhance JMAT modeling of patient holding capacity.** The number of holding beds needed at a treatment facility will vary and depend on evacuation availability. Modifications to JMAT would help planners more readily see this effect by distinguishing between patients who cannot yet be evacuated because they are recovering and patients who could be evacuated but are waiting for transport. Reporting the patient waiting times would allow the deterioration of patient condition to be modeled. (Patient deterioration would apply both to patients awaiting treatment and to patients awaiting transport to higher levels of care.) In addition, some patients, such as host country nationals, might not be evacuated. This would increase the need for holding capacity at the treatment facility and should therefore be included in planning tools.
- **Enhance JMAT modeling of patient evacuation.** The patient flow concept we describe illustrates how, in the current CONOPS, the treatment and evacuation of patients are closely linked. Because resources in one function affect the resources required in the other, it is necessary for a planning tool to include both in its modeling. We recommend that JMAT be modified to provide calculations of the numbers of aeromedical evacuation (AE) crews and critical care air transport (CCAT) teams that are needed to handle the patient flow, particularly on long-duration flights. In addition, JMAT 1.0.1.0 models AE by requiring the user to assign aircraft to locations. This is inconsistent with the current CONOPS, in which AE is carried out by aircraft of opportunity. Modifying JMAT to allow users to model airlift availability in terms of number and frequency of sorties would be more consistent with the CONOPS and would facilitate AE and CCAT crew calculations as well as calculations of patient holding requirements at treatment facilities.

Investments in these areas would greatly enhance the effectiveness of medical planning and, ultimately, enable commanders to deploy the right resources in the right areas. This would allow the whole system to be brought up to operational capability more quickly and would ensure that patient care and transport occur with a minimum of delay.

Acknowledgments

This work was sponsored by the Air Force Surgeon General, Lt Gen C. Bruce Green, and would not have been possible without his guidance and support. In the course of the project, we had numerous discussions with medical planners across the Air Force and in the joint community.[4] We especially thank Col Perry Cooper at the Air Staff; Maj Gen Douglas Robb, Col Brian Anderson, and Col Valerie Taylor at Air Mobility Command; CDR Debra Duncan at U.S. Pacific Command; Col Wayne Pritt and Lt Col N. Thomas Greenlee at 13th Air Force; Col Michele Williams and Col Laura Torres-Reyes at the Defense Medical Standardization Board; Lt Col Alan Murdock at Wilford Hall Medical Center; and Maj Joseph Lyons at the School of Aerospace Medicine. Many people helped us with securing access to and instruction on medical modeling software and data, and we wish to thank Paula Konoske and James Zouris at the Naval Health Research Center; Sharon Moser at the Office of the Secretary of Defense, Assistant Secretary for Health Affairs; Jim Whitcomb at Akimeka; and Johnny Brock and Sherry Adlich at Teledyne Brown Engineering, Inc.

---

[4] All ranks and positions were current as of the time of the research.

Abbreviations

AE	aeromedical evacuation
AF	Air Force
AR	Army
CASEST	casualty estimation
CASEVAC	casualty evacuation
CCAT	critical care air transport
COAA	course of action analysis
CONOPS	concept of operations
CONUS	continental United States
DNBI	disease or nonbattle injury
DoD	Department of Defense
EMEDS	expeditionary medical support
ER	emergency room
HUMRO	humanitarian relief operations
ICU	intensive care unit
JMAT	Joint Medical Analysis Tool
MCO	major combat operations
MTF	military treatment facility
NATO	North Atlantic Treaty Organization
NAV	Navy and Marine Corps
NHRC	Naval Health Research Center
OIF	Operation Iraqi Freedom
OR	operating room
PAR	population at risk

PCOF	patient condition occurrence frequency
STEP	stabilization, triage and treatment, and evacuation of patients
TML+	Tactical Medical Logistics+ model
TRAC2ES	U.S. Transportation Command Regulating and Command and Control Evacuation System
WIA	wounded in action

CHAPTER ONE

Introduction

The current concept of operations (CONOPS) for expeditionary medical care emphasizes getting patients to the highest level of medical care necessary as quickly as possible. Facilities in which this type of care is performed range from forward expeditionary units with minimal treatment capabilities to the most-advanced definitive care hospitals, all linked by medical transport capabilities. Executing this concept places a high premium on moving patients through a medical care system of increasing levels of care. The unifying theme across all these echelons and functions is the dynamic flow of patients through the medical system.

However, the measures currently used in the planning of expeditionary medical resources do not fully reflect this concept of patient flow. Rather than reflecting the dynamic flow of patients, they reflect more-static measures of capacity: The capability of medical treatment facilities is typically measured and expressed in terms of the number of hospital beds, and aero-medical evacuation capabilities are typically measured and expressed in terms of the number of aircraft and crews available. The measures used in planning are therefore not aligned with the CONOPS used in practice. Further, they do not lend themselves to the integration of planning across the treatment and evacuation functions or across the multiple levels of treatment facilities.

This report builds on an earlier study that proposed a new way of viewing expeditionary medical planning by defining a single metric for capabilities: patient flow rate.[1] This metric integrates and coordinates planning across echelons and functions and allows for a better understanding of the interdependencies among treatment and evacuation functions, medical capabilities within an echelon of care, and the military services.

In this report, we show how this new approach can be used within the Joint Staff's current medical planning tool, the Joint Medical Analysis Tool (JMAT),[2] to make planning for expeditionary medical operations more effective. In Chapter Two, we examine the current CONOPS for forward-deployed treatment and evacuation functions and then discuss how current planning processes do not reflect that CONOPS. In Chapter Three, we explain the principles that underpin the patient flow rate and then show how this construct can be applied to planning for forward-deployed treatment and evacuation resources. The proposed approach is applicable to the joint community, and it can be implemented using existing planning tools, with some modification, as outlined in Chapter Four. Chapter Five summarizes our conclusions and recommendations for the medical planning process, and particularly for JMAT.

---

[1] D. Snyder, E. W. Chan, J. J. Burks, M. A. Amouzegar, and A. C. Resnick, *How Should Air Force Expeditionary Medical Capabilities Be Expressed?* Santa Monica, Calif.: RAND Corporation, MG-785-AF, 2009.

[2] For more information, see Defense Health Information Management System, "Medical Analysis Tool (MAT) & Joint Medical Analysis Tool (JMAT)," factsheet, undated.

CHAPTER TWO

Air Force Medical Service Planning Is Not Aligned with the Current Concept of Operations for Patient Flow

The way in which treatment and evacuation functions are currently measured—and, thus, planned—is not fully aligned with the CONOPS currently employed by the military medical system. In this chapter, we provide background information on the current CONOPS and discuss how current metrics fall short in facilitating the integrated planning that is necessary to properly coordinate the different functions, services, and levels of care in the military medical system.

The Current Concept of Operations for Treatment and Evacuation

The current U.S. military medical CONOPS is built around the flow of patients between treatment facilities that provide increasing levels of patient care.[1] When service members are injured, the first level of care is provided by a *first responder (Level 1)*, who is the injured service member himself or herself (who administers self-aid), a nearby fellow service member (who administers buddy aid), or a medically trained responder (such as a medic or corpsman). Patients may then be brought to a *forward resuscitative surgical facility (Level 2)*, usually located close to the point of injury, where medical or surgical interventions may be performed to stabilize the patient for further transport. First-responder care and forward resuscitative care are generally the responsibility of the patient's military service. Forward treatment facility types vary with the different services, but, as different as these facilities are, they share certain characteristics: All receive patients who flow in from the point of injury or from a lower level of care, perform stabilizing treatment on them, and then flow them out to higher levels of care, if needed.

Patients who require more-advanced treatment may be evacuated to a *theater hospital (Level 3)*. Such facilities are typically deployed in theater and include modular tent setups and hospital ships. They provide increased surgical capabilities, a wider range of medical specialties, and greater inpatient capability. Examples of theater hospitals are the facilities at Bagram Air Base in Afghanistan and at Joint Base Balad in Iraq. At theater hospitals, patients receive the care necessary to allow them to return to duty or to be stabilized for transport to further care outside the theater. The theater hospital mission may be assigned to one or more services, and the hospital may see patients from any of the services.

---

[1] For descriptions of the terminology associated with different types of care, see U.S. Joint Chiefs of Staff, Joint Publication 4-02, *Health Service Support*, October 2006.

In recent years, there has been a tendency to evacuate patients out of the theater rather than to have them recuperate in the theater hospital prior to returning to duty. Outside of the theater, patients receive *definitive care*, which is the care necessary to "conclusively manage a patient's condition."[2] During operations in Afghanistan and Iraq, definitive care has been split into two categories: *outside the continental United States (Level 4)* and *inside the continental United States (Level 5)*. Level 4 and Level 5 patients have usually been evacuated to the Landstuhl Regional Medical Center in Germany, where they receive care before being brought back to hospitals inside the continental United States (CONUS). Figure 2.1 depicts an example medical network that illustrates the current CONOPS of the flow of patients from Air Force, Army, and Navy and Marine Corps populations between treatment facilities that represent increasing levels of care.

Linking all of these medical facilities is a variety of evacuation resources operated by the different services. These resources provide not only patient transport but also en-route medical care. Removing patients from the battlefield is called *casualty evacuation* (CASEVAC). Usually accomplished by a ground ambulance or by helicopter, CASEVAC brings the patient to forward resuscitative care (Level 2) or directly to the theater hospital (Level 3). CASEVAC in ground combat settings is primarily the responsibility of the Army or the Marine Corps, as determined by the combatant commander. Evacuation from Level 2 to higher levels is typically carried out by aeromedical evacuation (AE), which is generally understood to mean the evacuation of patients by fixed-wing aircraft. AE is typically the responsibility of the Air Force. Medical personnel accompany CASEVAC and AE missions to provide stabilizing care to patients.

Existing Metrics Are Capacity Based

Treatment Resources

Treatment resources in the Air Force exist primarily in the form of modular, tent-based facilities known as *expeditionary medical support* (EMEDS). Designed to support expeditionary air forces at forward locations, EMEDS can be built up as needed to provide increasing amounts of capability, including by increasing the equipment, personnel, and type of care they offer.

Figure 2.1
The Current Air Force Medical Service CONOPS for Treatment and Evacuation, Levels 2–5

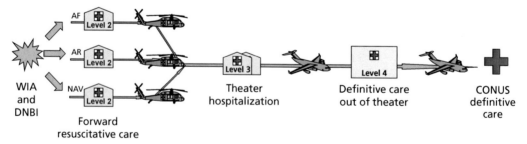

NOTE: AF = Air Force. AR = Army. DNBI = disease or nonbattle injury. NAV = Navy and Marine Corps. WIA = wounded in action.
RAND *TR1003-2.1*

---

2   Joint Publication 4-02, p. I-5.

EMEDS module names reflect the number of beds in the facility. For instance, EMEDS +10 is a module with ten beds, and EMEDS +25 is a module with 25 beds.

How big an EMEDS facility will be when it is deployed is decided after a review of published planning factors that assess the size of the population at risk (i.e., the number of service members at the deployed location) that the EMEDS will support. For instance, an EMEDS +10 is intended to support a population of 3,000–5,000.[3] The Air Force can also decide to deploy a mobile aeromedical staging facility or a contingency aeromedical staging facility. These are deployable facilities equipped and staffed to briefly hold or "stage" patients in preparation for transport. Like EMEDS, these facilities are categorized in terms of the number of beds they provide.

Evacuation Resources

Evacuation resources consist of ground or air transport vehicles and their patient-care crews. In the Air Force, these resources are fixed-wing aircraft and AE crews. The AE crews may be supplemented by critical care air transport (CCAT) teams in the case of patients who require more-intensive care during flight. (Other more-specialized teams, such as ones that care for burn patients, may be used as well.) In the past, certain aircraft were dedicated to the AE mission; one example is the C-9 Nightingale, a purpose-built airframe. With the C-9's retirement in 2005, AE has relied on "retrograde lift" and "aircraft of opportunity" provided by, for example, C-130 and C-17 cargo planes, and KC-135 tankers. Aircraft that have flown into an area while carrying personnel or supplies are rapidly reconfigured to carry patients and medical equipment on the return leg. In more-extreme cases, aircraft are diverted from their original mission and their cargo is deplaned in order to allow the aircraft to carry patients.

AE and CCAT crews are positioned at airlift hub locations and are transported as necessary to meet AE flights. Decisions about the number of crews to assign are, in theory, based on the number of missions that are expected to be flown. However, the lack of reliable casualty estimates, coupled with an understandable desire to not run short of capacity, can result in planning that is based on upper bounds. One Air Force interviewee noted that the number of AE crews assigned is based not on forecasts of AE demand, such as casualty forecasts, but rather on upper bounds on AE supply—namely, the number of cargo aircraft based at the location. (The logic was that, once crew rotations are taken into account, there is no need to assign more crews than the number of aircraft.) Although this anecdote does not supply conclusive evidence of practice, it does suggest the need for improved planning so that resources can be more efficiently used.

Disconnects Between the CONOPS and Planning Processes

Given the operational emphasis on patient flow, expressing treatment and evacuation capabilities in terms of such measures as numbers of beds and aircraft has two disadvantages. The first is that such measures are static measures of *capacity*: Beds and aircraft are fundamentally measures of the numbers of items. However, what is of concern to planners is not the number of items at each facility or function but rather the *capability* that can be provided by those resources. It is

---

[3] For more information about EMEDS, see U.S. Air Force School of Aerospace Medicine, *EMEDS Reference Handbook*, undated.

not readily apparent, for instance, whether a 25-bed hospital is sufficient to support a particular operational scenario; multiple factors—the rate at which patients become injured, the nature of the injuries, and the length of time that patients need to stay in the hospital—affect the resource requirements. In addition, these factors will vary from scenario to scenario.

The second disadvantage is that the treatment and evacuation functions use different units of measure. A CONOPS that requires a smooth flow of patients demands that the planning of treatment and evacuation be coordinated. With treatment resources being measured in beds and evacuation assets being measured in aircraft, it is difficult to know, for instance, whether one C-17 flight a day is sufficient to keep up with the number and mix of patients coming from a 50-bed hospital and whether a 25-bed treatment facility is sufficient to handle the patients who will be brought in by five helicopters.

In the next chapter, we propose a planning concept that is consistent with the military medical CONOPS and that helps integrate medical planning across treatment and evacuation functions, across the increasing levels of care, and across the different military services. This concept proposes that treatment and evacuation functions at all levels use patient flow rate (e.g., number of patients treated per day, number of patients transported per day) as the common unit of measurement for planning purposes. In our concept, treatment facilities are described not primarily in terms of the number of beds available (although that measure will always be useful) but rather in terms of the number of patients per day that can be seen at a facility. Likewise, evacuation resources are described in terms of the number of patients per day that can be transported. Together, treatment and evacuation resources are assigned on the basis of whether their patient flow rates are sufficient to keep up with the rate of injury and types of injuries and diseases that are expected to occur in a given scenario.

CHAPTER THREE

The STEP Rate: An Integrated Approach to Planning Treatment and Evacuation Resources

In this chapter, we propose a framework for tying together the planning of treatment and evacuation resources. The framework is based on the military medical CONOPS of patient flow. The goal of every treatment facility and every evacuation asset is the stabilization, triage and treatment, and evacuation of patients to the next and higher level of care as quickly as is prudent. Therefore, a measure of the capability of a component to provide care is the rate at which that component can carry out these activities. For evacuation resources, such as AE aircraft, the STEP rate would be the number of patients per day that can be transported. Such a measure would incorporate not only the number of aircraft assigned but also flying times and sortie rates. For treatment resources, such as field hospitals, the STEP rate would be the number of patients per day that can be treated and released or stabilized for transport. Such a measure would incorporate not only the rate at which patients arrive for treatment but also the types and severity of their injuries and diseases.

We propose that the patient flow rate, or STEP rate,[1] be applied across the entire medical network. This approach has the following advantages:

- **It corresponds to the CONOPS.** The expeditionary military medical system is built around the flow of patients away from the point of injury and toward facilities that can provide proper care. Measurement in terms of patient flow directly reflects this CONOPS.
- **It applies across all resources, levels of care, and the military services.** The measurement of patient flow can apply equally to ground- and ship-based treatment facilities and to different evacuation resources (e.g., ground ambulances, helicopters, fixed-wing aircraft). Further, the measure applies at any level of care (e.g., first responder, forward resuscitative care) and regardless of which service provides the capability.
- **It measures capability rather than capacity.** Rather than measuring resource levels (e.g., the number of beds, aircraft, and medical staff), the patient flow rate measures the capability provided by the resources. This enables planners to more readily assess whether a given set of resources can adequately support a given mission. The STEP rate applies not only to the individual assets providing treatment or evacuation functions but also to the medical system as a whole.

---

[1] RAND introduced the STEP rate in an earlier report (Snyder et al., 2009) that argues why the STEP rate is a more appropriate unit of measurement than beds and illustrates how the STEP rate would apply in different scenarios and configurations of the medical network. The authors suggest that treatment facility unit type codes be re-expressed in terms of STEP rates.

In this chapter, we explain the principles of the STEP rate and how it integrates the planning of the medical system across functions, echelons, and services. We then discuss how to use the STEP rate to determine what treatment, holding, and evacuation resources are needed to achieve a given patient flow.

The Principles of Patient Flow Rate

The CONOPS of patient care can be likened to the flow of fluid through a system of valves and holding tanks. In such a system, the rate at which fluid flows into and out of the holding tanks is governed by the valves. Ideally, the flow of fluid out of a holding tank is at least as fast as the flow of fluid coming into a tank; otherwise fluid backs up in the tank. Figure 3.1 illustrates this analogy.

In the figure, the flow of patients starts with casualties, including those wounded in action (WIA) as a result of combat and those suffering from diseases or nonbattle injury (DNBI).[2] Different scenarios will produce casualties at different rates and yield different mixes of injuries and illnesses. For instance, combat operations and humanitarian assistance efforts will produce vastly different types of patient conditions. Even within the category of combat operations, different phases of an operation—e.g., major combat versus stabilization—can result in a different mix of patients. The mix of injuries and illnesses can have a major effect on the patient flow rate that can be achieved with a given set of resources.

The flow of patients next progresses to the first treatment facility, a Level 2 facility for forward resuscitative surgical care. At this level, patients enter separate sets of facilities organized by service. Each facility is represented by two holding tanks. The contents of the first tank represent patients who arrive requiring stabilization or treatment; this tank can be thought of as the waiting room of a hospital emergency department. The first valve represents the rate at which patients at the facility are treated and stabilized for evacuation, and it governs the rate at

Figure 3.1
Analogy of Patient Flow in a System of Holding Tanks and Valves

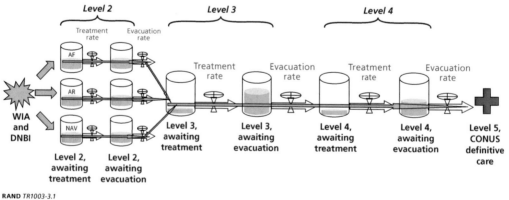

RAND *TR1003-3.1*

---

2   For simplicity, our figure omits Level 1, first-responder care, which may not represent an actual facility. Some services provide an additional level of care that is beyond first-responder care but still considered part of Level 1 (e.g., a battalion aid station or a shock trauma platoon). The general principle of measuring capability in terms of patient treatment rate could be applied to all of these treatment resources as well.

which the first tank is emptied. Patients who have been treated and are now stable enough for transport are moved to the second Level 2 holding tank, where they await evacuation to the next-higher level of care. This second tank corresponds to holding beds at the treatment facility or at an aeromedical staging facility. The rate at which patients leave this holding tank is governed by the rate at which the AE system (or some other transport resource) can transport them.[3]

The patient streams merge at a Level 3 treatment facility (i.e., theater hospital care), which may be a joint mission under the responsibility of one of the services, as designated by the combatant commander. From there, patients are evacuated by the Air Force to definitive care at Levels 4 and 5. At all levels, patients flow through holding tanks based on the rate at which they receive care and are transported to the next holding tank.

Although we have illustrated the flow as a sequential process with just a few facilities, actual operations may be more complex, involving multiple locations and multiple paths of flow. Each service is likely to operate multiple Level 2 facilities to support its deployed populations, and there would therefore be multiple points of entry into the medical system, with each point potentially having a different casualty rate. In a large operation, there could also be multiple Level 3 theater hospitals, which could be focused on supporting specific service populations. Patients could also be evacuated out of the theater into multiple Level 4 and Level 5 definitive care locations.

Further, the flow of patients may not be as sequential as the figure implies. For example, patients may not necessarily stop at every level in the system. Indeed, in current practice, some patients skip echelons. Patients evacuated from the battlefield may overfly a Level 2 facility and be taken directly to a Level 3 facility. Similarly, some patients who are severely injured, such as those with significant burns, may be evacuated directly from the Level 2 facility aboard extremely long flights, which require aerial refueling and specialized en-route care teams, to burn centers in CONUS (such as at San Antonio Military Medical Center, formerly Brooke Army Medical Center), thus bypassing Levels 3 and 4.

Some patients are not evacuated. The care received at Level 2 or Level 3 may be sufficient to allow them to return to duty, and such patients therefore do not require evacuation to higher levels of care. Conceptually, that situation could be represented by an additional exit valve after treatment at the location. Some patients are not evacuated for policy reasons. For example, it may make little sense to evacuate host country personnel (or civilians) back to CONUS. Also, in the case of a biological outbreak involving a contagious disease, it may be judged undesirable to move patients to another location and risk spreading the disease. In such cases, rather than moving to the next echelon of care, patients would follow different paths that lead to host country facilities or to alternative patient care or holding facilities. Even in these situations, the principles of flow rate apply.

Regardless of the specific path, the system still requires coordination between treatment and evacuation resources to ensure that the flow of patients is not unnecessarily delayed.

---

[3] Transportation is not continuous but rather usually occurs in discrete vehicle loads. Therefore, a more exact analogy might be a valve that periodically opens, releasing a measured amount of fluid each time.

The Role of the STEP Rate in Resource Planning

The goal of medical planning is to prevent resource imbalances that impede the rapid movement (as necessary) of patients across the entire medical system. The STEP rate proposed in this report will help planners by allowing them to better understand the interdependencies among treatment and evacuation functions, levels of care, and the military services. With this understanding, planners will be better able to assign resources in a way that ensures that these elements work together.

The STEP rate shows how shortages in resources—or the risk of shortages—in one area affect resources needed in other areas. For example, in the early stages of a conflict, the number of aircraft available for AE may be limited. Reductions in the number of available airlifters would reduce the evacuation rate between treatment facilities to a number below the rate needed to keep up with the flow of casualties. This reduced evacuation rate would mean that patients would have to wait longer at the lower-level facility. Consequently, planners would need to increase holding capacity at that location in the form of more holding beds (at a hospital or aeromedical staging facility), more medical personnel (to provide continued monitoring and care to these patients), and, possibly, additional medical interventions associated with the delay in getting the patients to higher levels of care.

Backups in patient flow at one level can have implications upstream. For example, if a Level 3 facility is filled to capacity, that could affect the ability to evacuate patients out of Level 2 facilities. Consequently, for reasons that have little to do with evacuation from Level 2 to Level 3 per se, planners would need to ensure that Level 2 facilities are equipped with more holding capacity.

These scenarios also illustrate the importance of coordination between the services. The Air Force, for example, needs to know the expected casualty stream from Army and Marine Corps units to plan for the right number of AE flights, crews, and patient movement items. Conversely, it is also vital that the Army and Navy medical planners are aware of Air Force AE capabilities because these capabilities could affect the holding capacity required at Army and Navy forward field hospitals and casualty receiving ships. The interdependence across functions and services also indicates that uncertainties—in the availability of AE, in the availability of treatment resources, or in the frequency and types of casualties—must also be shared across services, as they affect each others' planning.

The Importance of Casualty Forecasting

Casualty forecasting was not the focus of our study. However, any discussion of resource estimation must emphasize the importance of good casualty forecasts. If the estimates of casualties are inaccurate, the resulting estimates of medical resources required will be inaccurate, regardless of the quality of the resource estimation algorithm. As is the case with using any method for estimating medical resource requirements, using the patient flow rate or STEP rate concept requires an understanding of the number, type, severity, and timing of casualties in the scenario. The medical system's overall treatment and evacuation rate must be sufficient to keep up with the flow of patients entering the system. If the rate is not sufficient, the flow of patients will stall, and patients will face unacceptably long delays in waiting for treatment or evacuation.

The published planning factors for expeditionary medical support resources, such as EMEDS, associate a hospital of a particular size with an at-risk population of a particular size.[4] Such planning factors are implicitly based on assumptions about the casualty rates and injury mix that the population will suffer and about the treatment rate provided by a given hospital equipment set, given that mix of patients. When these assumptions are not made clear to the planner, a mismatch between resources and the particular scenario being planned may result, especially if the scenario the planner uses does not resemble the scenario assumed by the developer of the planning factors for the population-at-risk method. Tying resource estimates explicitly to estimates of casualty rates and patient mix is a more transparent method.

Casualty forecasts are therefore critical to proper medical planning. Different types of scenarios will lead to different casualty rates, casualty arrival patterns, and patient conditions (i.e., types and severity of injuries). For example, casualty forecasting tools built on the assumption that a large-scale armor attack will occur may not do well in predicting the rates and types of injuries that would be seen in a peacekeeping operation. Although no casualty forecast is likely to predict with certainty the events that will unfold in an actual contingency, efforts can and should be made to improve forecasting. We discuss casualty forecasting in more detail in Appendix A.

Determining Treatment Resources Needed to Achieve a Flow Rate

Medical treatment facilities must be sized and resourced sufficiently to keep up with the flow of patients coming in from the battlefield or from the previous echelon. Doing this requires translating a set of capacities—for example, the size of the emergency room, the number of operating rooms, and the number of medical personnel—into a measure of capability, such as patients treated per hour. To make this translation, we need to take a closer look at the elements within a treatment facility.

The concept of patient flow applies both between facilities and within them. In Figure 3.2, we present a simplified representation of a treatment facility. In this simplified facility, the major functions are the emergency room (ER), the operating room (OR), the intensive care unit (ICU), and the holding beds (or ward). Depending on their condition, patients either

Figure 3.2
Determining Treatment Resources Needed to Achieve a Flow Rate

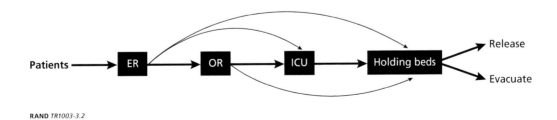

RAND TR1003-3.2

---

4 U.S. Air Force School of Aerospace Medicine, undated.

flow linearly through all the functions or skip certain functions; the figure's arrows represent potential paths. This level of detail may be sufficient, but, if more is required, additional steps, as well as ancillary and support functions (e.g., radiology), can be added.

Patients move from function to function within the facility—that is, from the ER to the OR to the ICU to the holding ward—as required. Each function has its own patient flow rate: The ER can triage and treat patients at a certain rate, the OR can perform surgeries at a certain rate, etc. Insufficient rates in one area can cause a backup in other areas. For example, if there are no holding beds available, as is often the case in civilian hospitals in the United States, patients cannot be moved out of the ER. The backlog of patients at the ER takes up space and resources that are needed to treat other patients, and this backup may eventually prevent the ER from seeing any new patients.

The patient flow rate that can be achieved by the facility is governed by the resources that are assigned to the various parts of the facility. Adding additional staff, equipment, and space to the ER would increase the number of patients per unit of time that the hospital can admit; similarly, adding more OR rooms and staff would increase the number of surgeries per unit of time that the hospital can perform.

Another factor affects the rate at which patients can be treated or stabilized at a facility: the mix of patient conditions (i.e., the types and severity of their injuries or diseases). Patient conditions play a role in what paths patients take through a facility; for example, some patients will not require surgery in the OR. Further, patients spend a different amount of time at each step, depending on their condition, because surgery and ICU recovery times differ based on the severity of the condition. It is therefore crucial to have accurate forecasts of the patient mix associated with a particular scenario.

With these points in mind, determining the flow rate that a facility can achieve can be accomplished in two ways. The first method is based on retrospective data from, or direct observations of, actual operations. Retrospective data from actual operations may be limited to aggregated numbers, such as the total number of patients seen during a 12-hour period of operation. Such data provide a very coarse estimate for the entire facility and therefore do not allow planners to adjust the resources pertaining to various elements within the facility. Patient treatment records could provide a finer-grained estimate of entry and exit times for each function within the treatment facility, but observations for every combination of conditions that a patient could have would be required and could well be unavailable. The amount of time a patient spends in the ER, OR, and so on could also be recorded through direct observations, but this could be labor intensive.

The second method is to use computer modeling. In this method, a computer program simulates the flow of patients through a facility, and the user varies the mix of patients according to the scenario being explored. This is the approach taken by users of the Joint Medical Analysis Tool (JMAT), the Joint Staff–approved tool for medical planning.[5] (JMAT is managed by the Office of the Secretary of Defense, Assistant Secretary for Health Affairs.) It is also the approach taken by users of the Tactical Medical Logistics+ model (TML+), which is a prod-

---

[5]   More information about different models can be found in Defense Health Information Management System, undated; Defense Health Information Management System, "Joint Medical Analysis Tool (JMAT)," web page, last updated August 22, 2011; and Theater Medical Information Program–Joint, "TMIP-J: A Portable Medical Information System," undated.

uct of the Naval Health Research Center (NHRC).[6] Computer models depend on assumptions about the care that is needed for each type of patient. The path taken by each patient, and the amount of time each patient spends at each point in the path, depend on the patient's conditions and must be specified in advance by clinical experts so that the information can be built into the model.[7] Ideally, this information is based on data from real operations.

Determining the Holding Resources Needed as a Function of Patient Arrival and Departure Rates

Some assets at treatment facilities serve more of a "holding" than a "treatment" function. For example, a ward bed occupied by a patient recovering from surgery can be thought of as serving a treatment function, in that this recovery time is part of the stabilization necessary before the patient can be transported. In contrast, a ward bed occupied by a patient waiting to be evacuated is serving a holding function, in that the patient is in that bed not because he or she is not yet fit for transport but rather because he or she is waiting for transport. Examples of assets that serve a holding function are a ward at a treatment facility, an aeromedical staging facility, and other types of medically supervised billeting arrangements.

In practice, this distinction between the treatment role and the holding role is less rigid. There is no obvious demarcation between when a patient is recovering and when he or she is simply waiting. The determination that a patient is well enough to be evacuated is based on the judgment of a physician, and it may also be driven by external factors beyond the patient's condition, including that space at the facility is limited, that there has been an influx of new patients, or that security conditions at the facility are deteriorating. These are examples of possible reasons why a patient might be released for transport earlier. In addition, patients who are being held still require treatment while they are waiting for transport. If transport to the next level of care is delayed, their condition may deteriorate, and they may require more treatment (including more surgeries) at their present location.

Nonetheless, the distinction between a holding function and treatment function is a useful one for modeling and planning purposes. The aforementioned caveats aside, the amount of holding resources required will depend on the availability of evacuation resources, but the amount of treatment resources required will not. If evacuation capability is limited, the treatment facility will require more holding resources, including those associated with providing medical treatment to patients who are held longer at the facility. Treatment facilities upstream could also require more holding resources. Conversely, if holding resources (e.g., medical staff, supplies, bed space) are limited, more evacuation resources may be needed. Given the possibility that AE could be constrained, it is especially important that planners understand both the trade-offs between holding resources and evacuation resources and the relative risks involved. Reducing the size of medical facilities requires an assurance of prompt evacuation, for instance.

---

[6] TML+ is discussed in Teledyne Brown Engineering, Inc., "Tactical Medical Logistics Planning Tool," user web portal, undated.

[7] At the time of this writing, the Defense Medical Standardization Board was, as part of its Common User Database project, convening panels to update these time estimates. JMAT 1.0.1.0, the version currently in use, employs older treatment data; the new Common User Database information is expected to be employed by JMAT 2.0.

Combatant commanders may not feel that the benefits of a reduced forward footprint are worth the risk of running short of treatment resources.

Further, there are scenarios in which evacuating patients out of the theater is not desirable. For example, it may make little sense to move injured host country personnel to CONUS or even to a third country far from their home. (This situation may arise especially frequently during humanitarian assistance and disaster response operations, where the bulk of the patients will not be U.S. service members but rather residents of the host country.) In such cases, a preferable course of action is to move the patient to definitive care at a facility within the country or, at least, within the region. Until that can be arranged, however, the treatment facility within the theater would have to be prepared to hold the patient longer and would therefore require more holding capacity. Planners should thus factor in this type of situation into their planning of facility capacity.

To better understand the trade-offs between holding and evacuation resources, planners need a planning tool that can calculate the required holding capacity as a function of the evacuation capability. In theory, planning by balancing patient flow rates makes calculating the trade-off relatively easy, as illustrated in Figure 3.3.

In the figure, we compare two systems, both of which see ten casualties per day and both of which have treatment facilities that can stabilize and treat (and thus make ready for evacuation) ten patients per day. In the system on the left, the evacuation resources are able to provide one flight per day that takes ten patients per day to higher levels of care. Therefore, in between flights, one day's worth of patients, or ten patients, build up in the holding area. In the system on the right, instead of one flight per day that takes ten patients at a time, the AE resources can provide one flight every other day that takes 20 patients at a time. This requires the holding resources to accommodate two days' worth of patients, or 20 patients. Note that planning factors that associate a facility of a particular size with an at-risk population of a particular size (e.g., an EMEDS +10 corresponds to a population of 3,000–5,000) do so without regard to the frequency and size of evacuation flights.

There is certainly room to refine this calculation further by incorporating random variation in the casualty, treatment, and evacuation rates. Queuing models and computer simulations are the logical tools to use to make these calculations.[8] Regardless of this variation, the

Figure 3.3
Two Illustrative Examples of Calculating Holding Capacity

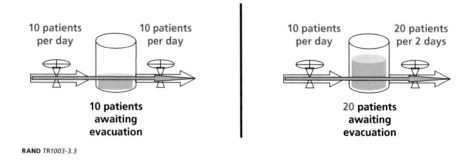

---

[8]   Currently, JMAT allows the user to select stochastic variation in the creation of casualties. However, it is not stochastic beyond that point, including within treatment routines, treatment times, evacuation times and priorities, and routes.

same conclusion applies in that calculating holding resource requirements is relatively straight-forward under the concept of patient flow.

Existing planning tools, such as JMAT, are close to being able to perform this calculation. However, JMAT requires users to express AE in terms of numbers of aircraft assigned to particular locations and routes. It is more difficult to see an imbalance between treatment and evacuation when it is not readily apparent how a given number of aircraft translates into a number of patients evacuated per day. Expressing evacuation capability in terms of patient flow—such as by converting airframes, distances, and sortie rates into readily comprehensible numbers of patients evacuated per day—would make the effect of imbalances more apparent. Recommendations for modifying JMAT to better handle the calculation of holding resources are presented in detail in Chapter Four.

Determining the Evacuation Resources Needed to Achieve a Flow Rate

Under the STEP rate concept, the unit of measurement for evacuation is the number of patients evacuated per unit of time. (A more detailed version of this measure would break the number of patients into different types: the number of ambulatory, litter, and critical patients, for instance, evacuated per unit of time.) Although evacuation resources are often measured in terms of the number of ambulances, helicopters, or aircraft, the STEP rate concept more readily translates into the sortie rate. Further, expressing the requirement for AE in terms of flow is aligned with the current CONOPS of using retrograde lift and aircraft of opportunity (rather than requiring dedicated airframes).

Planning evacuation thus amounts to determining the number of sorties required to sustain the necessary patient flow rate. Once the number of sorties is determined, it should be relatively straightforward to translate that number back into the number of airframes, air crews, and AE and CCAT crews required. Planning factors from Air Mobility Command provide both the number of patients who can be carried by various types of aircraft and the AE and CCAT crew ratios required.[9]

However, the evacuation mission can be complicated in some scenarios due to the distance and, especially, the flying time involved in evacuating patients between facilities. In current operations in Afghanistan and Iraq, patients are evacuated from the theater hospitals in those countries to Germany, where they receive Level 4 treatment at the Landstuhl Regional Medical Center. This trip is approximately 2,800 nautical miles, which, with the overflight permissions that have been granted, translates into a flying time of approximately seven hours. Interviewees at Air Mobility Command indicated that recent experience with caring for patients has led to a preference that evacuation flights last no longer than seven hours. Therefore, we use this distance as an estimate of the distance that patients can be safely flown and, consequently, as an estimate for how far the Level 4 hospital should be from the Level 3 facilities.

---

[9] See, for instance, U.S. Air Force, Air Force Pamphlet 10-1403, *Air Mobility Planning Factors*, December 18, 2003; U.S. Air Force, Air Force Tactics, Techniques, and Procedures 3-42.5, *Aeromedical Evacuation*, November 2003; U.S. Air Force, Air Force Instruction 11-2AE, *Aeromedical Evacuation Operations Procedures*, May 18, 2005a; and U.S. Air Force, Air Force Instruction 11-2AE, *Aeromedical Evacuation Operations Procedures*, Vol. 3, Addenda A: *Aeromedical Evacuation Operations Configuration/Mission Planning*, May 27, 2005b.

Figure 3.4
Areas Within Reach of Level 4 Facilities

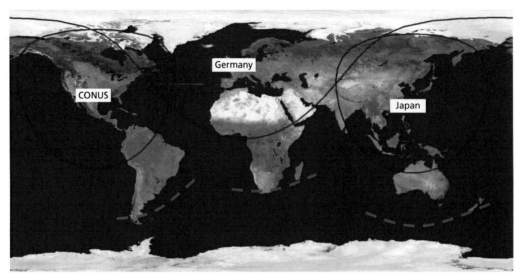

RAND *TR1003-3.4*

The solid curves in Figure 3.4 indicate the areas around the world that are within 2,800 nautical miles of established U.S. military medical infrastructure in the United States, Germany, and Japan that can provide Level 4 care. These curves show that much of the world, including nearly all points in the northern hemisphere, is within direct AE reach of a preferred Level 4 facility, assuming that overflight permissions can be obtained.

However, future scenarios could require military medical support in areas that lie beyond reach, such as South America, Australia, the India-Pakistan border area, and the southern parts of Africa. If so, the military medical system would have essentially two options: fly AE missions of longer duration or add stopover locations. Either option has resource implications.

Extending Aeromedical Evacuation Flight Times

Extending the flight distance from 2,800 nautical miles to, for instance, 5,000 nautical miles would put the entire populated world within reach of three Level 4 facilities, as shown by the dashed curves in Figure 3.4. Routes of this length would extend flying times to approximately 12 hours, although overflight restrictions could easily extend the flying time to 15 hours. Longer flying times could require aerial refueling, which could be a burden on tankers.[10] Of more direct concern to medical planners is the burden that longer flying times would place on patients and on AE crews.

Interviews with Air Force medical experts indicate that there is no universally accepted maximum flying time for patients, but, as noted earlier, experience in U.S. Central Command has led to a preference for flying times of seven hours or less. The stresses associated with flying can be detrimental to patients, and, as a general rule, no patient gets better in the air. However,

---

[10] The range of a cargo aircraft depends on its load. The manufacturer of the C-17 indicates that the aircraft's range with a "paratroop" load (i.e., passengers but no cargo) of 40,000 lbs is 5,610 nautical miles (Boeing Defense, Space & Security, "C-17 Globemaster III," Boeing backgrounder, May 2008). This 40,000-lb "paratroop" configuration may be similar to conditions during AE. It is therefore possible that C-17s could perform 5,000–nautical mile AE flights without aerial refueling.

more research is needed to understand the effects of flight on various patient conditions and to be able to assess the costs and benefits of longer flights.

Longer flights would also increase requirements for the number of AE crews and CCAT teams. The maximum crew duty day for an AE crew is 18 hours. If the length of the mission were to cause the duty day to exceed 18 hours, additional personnel would be needed to augment the crew. Adding more crew members to a flight reduces the amount of space available for patients, thus affecting the capacity of each flight. Further, additional treatment might be required at treatment and holding facilities on the ground before a patient is cleared for a longer flight.

Adding Stopover Locations

AE range can also be increased by adding stopover locations between the in-theater Level 3 facility and the out-of-theater Level 4 facility. This approach eliminates the need for aerial refueling and some other problems associated with extending flight distances, but it introduces other complications. For example, additional treatment facilities may have to be deployed. Resource estimation could be accomplished in the same way used for other levels of treatment facilities, but planners would require guidance on the care that is needed by a patient at a stopover location. Interviews with Air Force physicians indicate that the resources could range from a simple aeromedical staging facility to a full Level 3 theater hospital (so that a patient never goes from a location of higher care to a location of lower care).

The deployment of medical resources could be reduced if local medical facilities could be leveraged, but information on the resources available at these locations is limited. Surveys of a handful of locations aligned with likely U.S. Transportation Command air hubs would facilitate planning. Even so, access to these facilities would be far from guaranteed, and using them could be undesirable for political or force protection reasons. Consequently, planners and programmers may have to plan for the possibility of deploying additional medical resources to serve as stopover locations. Additional considerations in selecting stopover locations are discussed in Appendix C.

CHAPTER FOUR

Recommendations for JMAT

To implement the integrated, patient flow concept of planning described in Chapter Three, planners need tools to help them estimate the resources required to achieve the desired STEP rates for all of the treatment facilities and for the evacuation functions that link them. Fortunately, existing planning tools already perform many of these functions. Modifications in a few areas would fully enable the tools to model the integration between treatment and evacuation, thereby helping planners estimate the resources required at each function and location and weigh any trade-offs that may result. In this chapter, we identify these modifications.

The tool approved by the Joint Staff for use by medical planners in developing the medical section of their operations plans is the Joint Medical Analysis Tool (JMAT). JMAT is a Windows-based military medical stochastic simulation tool used by medical planners to generate theater medical support requirements. Using JMAT, military medical planners determine the level and scope of medical support needed for a joint operation. The system was developed by the U.S. Theater Medical Information Program within the Office of the Secretary of Defense, Assistant Secretary for Health Affairs, and the Joint Staff/J4/Health Service Support Division. The original contractor, Booz Allen Hamilton, produced the model as the Medical Analysis Tool, and, in 2006, a new contractor, Akimeka, took over as developer. As of September 2010, the current version is JMAT 1.0.1.0 (often referred to as "version 10-10"). A new, web-based, collaborative planning version, JMAT 2.0, is currently under development,[1] but it was not yet ready when we conducted our study.

Another, similar tool is the Tactical Medical Logistics+ model (TML+), which was developed by the Naval Health Research Center (NHRC) and Teledyne Brown Engineering, Inc., and is managed by NHRC. TML+ focuses on the lower echelons of care (Levels 1 and 2) and, compared with JMAT, adds more details in many areas, including a probabilistic died-of-wounds algorithm. The capability to model patient deterioration is also being added to TML+.

In this report, we have focused our attention on JMAT because that is the model used by the Air Force and by combatant command planners. Some of our recommendations are specific to JMAT, but many of the principles apply to planning regardless of the tool being used.

---

[1] Defense Health Information Management System, undated. As noted earlier, JMAT 2.0 will include, among other things, updated patient treatment information that is being compiled by the Defense Medical Standardization Board.

Overview of the JMAT Model

JMAT's purpose is to help medical planners calculate the amount and type of medical resources that are required for an operational scenario. It does this primarily by simulating the scenario and reporting whether the allocated resources are sufficient to meet the needs of the operation.

JMAT allows users to develop a medical network and to model the interaction of processes that would occur in a theater of operations. Given casualty sources, daily rates of battle casualties (derived from operational plans), and DNBI rates (derived from operational experience), JMAT generates a stream of patients with specific injuries and illnesses and tracks their progression through a doctrinally correct medical treatment and evacuation system. For a given scenario, the model runs a simulation for a specified period to generate casualties by day and by individual patient condition for more than 300 patient conditions. JMAT processes the patients in the medical system by treating them based on guidelines specific to each patient condition and by processing and evacuating them through the different levels of medical care.

JMAT can be run in two different modes. The first, the requirements estimator mode, yields a quick but coarse estimation of resource requirements based on average rates. It starts with a user-specified population at risk (PAR) and the percentage of those individuals that become casualties throughout the course of the scenario.[2] Using that information, JMAT determines the number of patients who arrive at each medical facility and subtracts from that the number who die in a hospital. A rate of evacuation to the next echelon of care is then applied to those survivors, and the model then returns the remaining casualties to duty. Based on this patient flow, JMAT reports the number of beds, staff, supplies, and evacuation assets, by level of care and by day, that are required to treat the injured population. If the user has also defined a medical network, JMAT reports shortages and overages based on the network's capabilities and on the resources needed to treat the injured population.

The second mode, the course of action analysis (COAA), provides a much more detailed calculation that is based on a stochastic simulation.[3] Whereas the requirements estimator treats all patients identically and applies average rates across the board, the COAA disaggregates the casualties into patients with different conditions in accordance with the patient condition occurrence frequency (PCOF) file that is built into the tool. The program then simulates the treatment of each patient at each facility according to the information built into time-task-treater files. JMAT then produces detailed reports of the usage and shortages of beds, staff, supplies, and evacuations, by day and by specific facility, and of how much time individual patients spend at each facility.

The next section describes JMAT's inputs and outputs and, where relevant, refers to COAA estimation.

---

[2] The requirements estimator does not make use of the patient condition occurrence frequency to divide casualties into specific types. Rather, it treats an aggregate number of casualties.

[3] As mentioned earlier, the stochastic component is available only at the point of patient-stream creation. Treatment, evacuation, and routing cannot be done stochastically.

JMAT Inputs

To determine the medical requirements needed in theater, the medical planner must enter a great deal of information about operations, casualties, and medical capabilities into the model. The basic scenario details required include climate (e.g., North Atlantic Treaty Organization [NATO] locations are temperate, Northeast Asian locations are cold), scenario duration (in days), and, if appropriate, a casualty rate set (consisting of notional rates, NATO rates, or user-defined rates). Evacuation policy determines the maximum number of days a patient can stay at an echelon of care before being evacuated to the next-higher level.

Casualties

After basic setup, the first details the medical planner must add to a scenario concern the casualties that will flow through the medical network. The user inputs information about the different sets of people who may suffer casualties, such as military units at specific locations. For each of these populations, the user specifies the PAR, the day of arrival, daily additions and subtractions to the combat and support populations, and the rate at which the population is replaced as casualties are incurred.

There are 12 different types of casualties that can affect a given population: wounded in action, killed in action, missing in action, captured, administrative, battle fatigue, disease, nonbattle injury, nuclear, biological, chemical, and outpatient visit. The user specifies casualty rates (in terms of casualties per 1,000 troops per day), with separate rates for combat and support troops. For four of these casualty types—wounded in action, nonbattle injury, disease, and battle fatigue—casualties are further broken out into more than 300 possible patient conditions.[4] The frequency distribution of these conditions is governed by the PCOF file; for each population, users can select from several PCOFs built into JMAT, or they can input their own.

When the user runs JMAT, the simulation generates for each population a random stream of patients. The patients arrive at a rate according to the casualty rates specified by the user, and the injuries or illnesses are distributed across patient conditions in accordance with the PCOF table selected or input by the user.

The Medical Network

Once casualties are generated using the PAR and rates of occurrence, patients flow through a medical network. First, each patient arrives at a Level 2 facility specified by the planner; then, he or she is moved through Levels 3–5 until he or she is discharged or dies in a hospital. JMAT has 31 built-in U.S. medical facilities; 11 are Level 2, 12 are Level 3, seven are Level 4, and one is Level 5. There are also NATO facilities provided by Belgium, Denmark, Germany, Italy, Norway, and the United Kingdom. Once the user defines the point of entry for each casualty source, the distances between facilities, and how those factors are connected (i.e., who goes where), the user specifies the resources available at each facility.

---

[4] The current list of patient conditions and corresponding treatment information was inherited by JMAT from its predecessor program, the Medical Analysis Tool. The Defense Medical Standardization Board is updating those files to include more than 300 patient conditions, coded according to the International Classification of Diseases, and their accompanying time-task-treater information. The Office of the Secretary of Defense has tasked NHRC with developing new PCOFs for all types of scenarios, including humanitarian assistance and disaster relief. These updated patient frequencies will be included in JMAT 2.0. Note that planners can create their own user-defined PCOF.

The medical planner specifies the day a medical facility becomes available and operational, and it remains in this state throughout the duration of the scenario. The number of beds,[5] staff,[6] supplies,[7] and evacuation assets[8] are defined by their daily availability.

Evacuation Assets

Evacuation assets, including aircraft, are specified in JMAT in one of two ways. They may be assigned either to a medical facility or to a separate "bed-down" facility (i.e., some other air base location that is not associated with a medical facility).

In JMAT, evacuation assets that are assigned to a medical facility are owned by the lower level of care, and they make trips only to transport patients to the next level. For example, a C-17 that is assigned to a Level 3 facility takes patients who are ready to be evacuated from that facility to an assigned Level 4 facility and then returns to its home at the Level 3 facility. JMAT limits each aircraft to one trip per day, although the total number of aircraft available can be varied across days.

In JMAT, evacuation assets that are assigned to a bed-down facility are not tied to a specific medical treatment facility. However, they do run a specified route that starts at an entry point and has them move patients to one or more facilities in the network. The user defines the type, capacity (i.e., number of ambulatory and litter spaces available), and number of aircraft at a bed-down facility. The user also defines the turnaround time between legs of a route and the flight plan, including flight time and distance between facilities. When aircraft are assigned in this way, the number of aircraft available remains constant throughout the duration of the scenario and cannot be changed. In other words, if a C-130 bed-down facility with two aircraft becomes available on the fifth day, those two assets will run the designated route once per day for the remaining days in the scenario.

JMAT Outputs

After running multiple iterations of the scenario, JMAT generates information on the assets available and used. This information reveals shortages and overages in the medical network.

Specifically, JMAT reports, by day, by medical facility, and by bed type, the number of beds and bed hours available, the number of hours each bed was used, and the percentage of available time each bed was used. The same type of information is provided for each of the 14 different staff types, and JMAT also reports, by day and facility, the quantity of blood and class VIIIA medical supplies available and used.

JMAT provides similar information for evacuation assets assigned to bed-down facilities. For each asset at each bed-down facility, JMAT reports, by day, the number of ambulatory

---

[5] The following bed types are built into JMAT: emergency room, clinic, operating room, intensive care unit, intermediate care ward, minimal care ward, convalescent care ward, and staging.

[6] The following staff asset types are built into JMAT: anesthesiologist, clinical enlisted, internal medicine, general/thoracic surgeon, medical support officer, medical surgical nurse, neurosurgeon, obstetrician/gynecologist/urologist, ophthalmologist, operating room nurse, oral surgeon, orthopedic surgeon, other medical officer, and radiology/lab.

[7] JMAT has two types of supply assets: blood and class VIIIA. Blood is measured in units, and class VIIIA supplies are generically measured in pounds.

[8] JMAT includes 40 U.S. and 31 NATO built-in evacuation assets.

and litter spaces available and used. This provides some information on the mix of patients being transported.[9] Regardless of whether evacuation assets are assigned to medical facilities or to bed-down facilities, if there are patients who are waiting for evacuation due to shortages, JMAT reports the reason for that shortage (e.g., lack of staff, supplies, beds, or evacuation assets).[10]

Finally, JMAT outputs information on patient movement. By facility and by day, the model reports the number of patients admitted from combat (for Level 2 facilities, it reports only nonzero numbers), admitted from evacuation (for Levels 3–5, it reports only nonzero numbers), in treatment, waiting for treatment, evacuated, waiting for evacuation, in transit, returned to duty, and who die in a hospital. More-detailed information is available concerning the types of conditions of the patients arriving at each facility on a given day, and, for those same conditions, the model reports, by day, the number who are treated, who return to duty, and who die.[11] The most-detailed information is contained in the patient movement report, which, for each individual patient, details when the patient arrives at a facility, how long he or she spends in each treatment area, how much time he or she spends waiting for evacuation, and how much time it takes to evacuate him or her from one echelon of care to the next.

Recommended Modifications to JMAT

Because JMAT simulates the flow of patients within and across facilities, it already operates in a manner consistent with the patient flow rate concept described in Chapter Three, even though it does not explicitly express treatment or evacuation capabilities in terms of the STEP rate. JMAT appears sufficient in the area of estimating patient treatment resources at Level 3, 4, and 5 facilities; work being conducted by the Defense Medical Standardization Board to revise the time-task-treater files that will be used in JMAT 2.0 will enhance the accuracy of the modeling of patient treatment.

However, to more fully implement the STEP concept, there are three main areas of JMAT that would require some modification. We recommend the following:

1. Extend JMAT to include the entire chain of care.
2. Enhance JMAT modeling of patient holding capacity.
3. Enhance JMAT modeling of patient evacuation.

Recommendation 1: Extend JMAT to Include the Entire Chain of Care

The flow concept we present in this report emphasizes the interdependence among treatment and evacuation functions and among treatment facilities in different echelons. This interdependence requires a planning tool that encompasses all levels of care. At minimum, the

---

[9] Information on the number of AE and CCAT crews, or on the number of critical and noncritical patients, would improve the user's ability to plan for a given scenario.

[10] Transport delays due to evacuation shortages can be caused by the following reasons: there are no evacuation assets, the required assets are unavailable, no routes are available, the asset is not in range or is road restricted, fixed wing/rail capacity is full, there are no free assets, the asset is restricted due to route logic, or there is no room at the bed-down facility.

[11] Not all of the more than 300 patient conditions are reported. Rather, similar conditions are grouped into a total of 79 injury categories.

tool should include the path of the patient from the point of injury to forward resuscitative care (Level 2) through arrival in the United States for definitive care (Level 5), as applicable. Current planning tools fall short in the three areas discussed in the remainder of this subsection.

Modeling Level 2 Care

A shortage of evacuation capability to Level 3 can affect the holding resources necessary at Level 2, so including Level 2 care in the model is important for proper planning. It is particularly crucial because the transition from Level 2 to Level 3 care is often the point at which responsibility for the care of a patient moves from one service to another, and care across the seams between service boundaries may require coordination. However, the version of JMAT available as of September 2010, JMAT 1.0.1.0, does not include care received at Levels 1 and 2 in its simulation modeling.[12] Instead, it models treatment resources starting at Level 3, effectively assuming that all patients from Level 2 proceed to Level 3. Thus, by not including full modeling of Level 2 care, JMAT does not provide a good estimate of the resources required at Level 2, and it may overestimate the resources required at Level 3 and beyond. Carrying this argument further, one may wish to model the entire chain of care—from the point of injury onward—and, for completeness, include even Level 1 care.

We wish to note that TML+ does model Level 2 care, including the resources and transportation assets required at a Level 2 facility to handle a given patient stream and the effect of the distances between treatment facilities on patient treatment. Indeed, it models these requirements for Level 3 as well (including EMEDS and some surgical specialty care). One solution may be to couple TML+ with JMAT. However, differences in each model's assumptions and modeling algorithms would have to be reconciled.

Providing for the Possibility of Stopover Locations

As discussed in Chapter Three, future operations may occur in locations that are far from existing Level 4 facilities. The greater distances, coupled with possible restrictions in overflight rights, could mean that flights between Level 3 treatment and Level 4 treatment are unsatisfactorily long. Stopover locations between Levels 3 and 4 may therefore be required.

JMAT does not currently provide a means for modeling stopover locations. It allows the user to add a military treatment facility (MTF) that serves as a "pass-through facility," but patients do not receive any treatment at these pass-through locations. Instead, their normal course of treatment is deferred until they arrive at the next MTF. Thus, JMAT as currently configured would not properly model the care that might be required at a stopover location.

Adding capability to JMAT to include stopover locations would require the establishment of additional medical doctrine. Interviews with Air Force surgeons at Wilford Hall Medical Center indicate that there is no clear guidance on the care that would be necessary at such a stopover location. In concept, the stopover location might be as simple as an aeromedical staging facility, if one could assume that the patient would not require much medical care during the stopover. This assumption is likely unsafe, however. Medical professionals prefer that a patient never go from one level of care to a lower level: A patient leaving a Level 3 facility, for example, should be moved at least to another Level 3 facility, if not a higher-level facility. Ideally, then, although a stopover location is not intended to serve as the point at which further

---

[12] JMAT does include some rudimentary Level 2 modeling in its "requirements estimator," which provides a quick estimate of beds, staff, supplies, and evacuation assets based on simple planning factors only. Unlike the full COAA simulation, the requirements estimator ignores the different patient types that would be associated with different scenarios.

treatment is provided, it should have the capability to take care of any patient who develops complications en route to that location.

After the necessary medical doctrine is established, time-task-treater files would have to be established for care at a stopover location, including the amount time spent by a patient at the location, the procedures that may need to be undertaken, and the resources that would be required, all expressed as a function of the patient's condition. These would serve as JMAT's input parameters for modeling the requirements necessary to maintain the flow of patients through the stopover facility.

Allowing Echelons to Be Skipped

Although we have thus far emphasized the need to include in the model every echelon through which a patient might travel, not every patient will go through every level of care. For example, further evacuation is not needed in all cases. Some patients can be treated at their current location and therefore do not require transport to higher levels of care; JMAT includes provisions for this situation (with the notable exception of Level 2, as discussed earlier).

When patients do require evacuation to higher levels of care, their progression through the levels is not necessarily sequential. In current operations in Southwest Asia, for example, patients are sometimes brought directly from the point of injury to the Level 3 theater hospital facility, bypassing Level 2 care. Similarly, patients sometimes bypass Level 3 care in theater and are flown directly from Level 2 to Level 4 care in Germany or to Level 5 care in the United States. Such decisions are often based on the severity of the patient's injuries. For example, patients with severe burn injuries need to be treated at a specialized burn center but must also be transported within a limited window of time if they are being transported via aircraft. Decisions to have a patient bypass a level of care can also be based on the availability of treatment and evacuation resources at the various locations. For example, an inability to evacuate patients to Level 3 because of lack of capacity may result in transport to Level 4 or Level 5. Skipping echelons is compatible with the patient flow rate concept described earlier; it is simply an alternative path of flow. JMAT, however, only allows patients to skip Level 4 care. (The user specifies some fixed percentage of patients who are evacuated from a Level 3 facility directly to a Level 5 facility.)

Recommendation 2: Enhance JMAT Modeling of Patient Holding Capacity

Some of the resources at a treatment facility, such as the ER and the OR, can be assumed not to vary with outbound evacuation capacity: The treatment resources are necessary for keeping up with the flow of patients entering the facility, and they support the necessary function of stabilizing the patients before they can be evacuated. However, other resources, such as the number of ward beds (and, perhaps, ICU beds, since some patients being held would stay in the ICU rather than in the ward before being evacuated), vary somewhat with outbound evacuation capacity: If a patient requires evacuation to a higher level of care, once he or she is stabilized enough for transport, the only reason why he or she would remain at the facility is because he or she is waiting for evacuation. If evacuation capacity is limited, the need for ward beds and ICU beds would increase.

Of course, some amount of ICU and ward capacity would be needed even if evacuation were plentiful; this is because, in general, patients need time to recover before they can be transported. Ward beds and ICU beds thus serve two purposes: a treatment purpose, where the beds play a role in stabilizing patients before evacuation, and a holding purpose, where the beds accommodate stabilized patients as they await evacuation. Current planning tools fall short in the three areas discussed in the remainder of this subsection.

Distinguishing Treatment and Holding Functions

Planning tools, such as JMAT, should allow the planner to see both (1) the distinction between the treatment (or recovery) role and the holding role played by ICU and ward beds and (2) how the holding capacity responds to changes in evacuation capacity. The treatment briefs within the Common User Database may need to be revised so that they can delineate when patients within the ward and ICU are considered fit for transport.

Reporting Patient Waiting Numbers and Times

Distinguishing the treatment role from the holding role makes it easier to report the number of patients awaiting treatment or awaiting transport and to report their waiting times. This information would enable planners to see the effects of shortages in treatment and evacuation capability. It would also enable further modeling of (1) whether and how much patient conditions deteriorate as patients await treatment or transport and (2) how that deterioration may lead to a need for additional treatment. With these modifications, the modeling tool could show the effects of delays not just on resource requirements but also on health outcomes.

Modeling Nonevacuating Patients

In the case of major combat operations, JMAT is designed to model the medical resources needed for situations in which patients who are not returning to duty will eventually be evacuated to CONUS. However, in other types of scenarios, patients may not be evacuated from the point of injury. For example, in the case of chemical, biological, radiological, or nuclear warfare, patients would remain in theater, perhaps at the facility at which they first entered the medical system, to limit the risk of spreading contamination (or contagious infection) to other areas. Depending on the resources in theater, patients in that facility might receive treatment normally associated with higher levels of care. Similarly, when host country patients are being treated, it may not make sense to evacuate them away from their home countries; however, there may also not be appropriate host country facilities to which such patients can be safely transferred.[13] A more flexible planning tool would accommodate both situations: the one in which service members are eventually returned to CONUS for definitive care and the one in which patients are treated at the point of injury rather than being moved.

Recommendation 3: Enhance JMAT Modeling of Patient Evacuation

The patient flow concept described in Chapter Three illustrates how, in the current CONOPS, the treatment and evacuation of patients are closely linked. Because resources in one function affect the resources required in the other, a planning tool must include both functions in its modeling. To ensure that patient evacuation methods are consistent with the current CONOPS, JMAT requires modification in the three areas discussed in the remainder of this subsection.

Aligning AE Modeling with the CONOPS

JMAT handles AE in a manner that is not consistent with the current CONOPS of AE operations and planning. This makes it difficult for users to see how changes in AE affect

---

[13] Air Force medical personnel report, for instance, that most of the longer-term occupants of theater hospital beds in Iraq are Iraqi patients, since American patients are typically evacuated to Germany and CONUS.

requirements for evacuation resources and treatment facility holding resources. JMAT currently requires the user to specify evacuation resources in one of two ways. The first way is for aircraft to be assigned to a treatment facility location. Aircraft at that location serve the purpose of evacuating patients to the next-higher level of care (e.g., an aircraft assigned to a Level 3 facility takes patients to a Level 4 facility); it is assumed that each aircraft makes one trip per day. The second way in which JMAT allows users to specify evacuation is to have the aircraft assigned to a bed-down facility from which the aircraft flies a particular route that the user specifies; again, the assumption is that each aircraft makes one trip per day.[14]

Both of these methods are inconsistent with the current CONOPS for AE. In the current CONOPS, aircraft are not necessarily dedicated to the AE mission, and they are neither assigned to treatment facilities nor necessarily assigned to fly fixed AE routes. Rather, AE is carried out by aircraft of opportunity: Aircraft that fly into a location carrying personnel or supplies are, as needed, rapidly reconfigured to carry patients on the departing trip. Aircraft can also be diverted from their existing route (and their cargo can be offloaded) to fly an AE mission instead. Consequently, the parameter of interest to planners who are seeking to weigh the trade-off between patient holding resources at a treatment facility on the one hand and evacuation resources on the other is not numbers of aircraft but numbers of sorties.

JMAT should therefore be modified to allow users to vary airlift availability by specifying the number and frequency of sorties. In addition to the benefit of aligning JMAT with the CONOPS for AE, expressing AE in terms of sorties per day (or, alternatively, days between sorties) facilitates the computation of the STEP rate for evacuation, which in turn facilitates the calculation of the patient holding requirements at treatment facilities. Ideally, the user will be able to vary the sortie frequency during the course of the campaign being simulated.[15]

Calculating the Number of AE Crews

Evacuating patients requires not only aircraft sorties but also accompanying patient care crews. A lack of crews could hamper evacuation rates, causing a buildup of patients at treatment facilities. However, JMAT does not currently report the number of AE crews needed. Calculating the number of AE crews needed should be straightforward once the number of sorties is known. Planning factors related to the length of time necessary for a crew to return to the originating location and related to the amount of crew rest need to be developed and incorporated into JMAT.

In addition, the distance and, particularly, the flying time between facilities must be taken into account. Long flights, such as those associated with operations located far from existing medical infrastructure, could require an augmentation of AE crews due to the duty-day restrictions that apply to all flight crews. JMAT 2.0, which will have a geographic information system interface, is expected to model distances more capably than JMAT 1.0.1.0, and this will facilitate the computation of the number of AE crews needed, especially for longer flights.

---

[14] As noted earlier, JMAT assumes that the number aircraft assigned to a bed-down facility is static through the entire scenario, although the number of aircraft available can vary day to day when the aircraft are assigned to an MTF.

[15] Note that TML+ allows the user to define pools of transportation assets that are not attached to a treatment facility. An unattached asset travels to the first patient needing transport, regardless of the patient's location. Although the user cannot adjust the sortie rate, TML+ does allow the user to determine at what times during the scenario each asset is available, and availability can vary throughout the scenario. The user can also define the patient loading time for each asset and how much time passes between when a request for transport is made and when the asset is ready to travel.

Calculating the Number of CCAT Crews

In addition to the AE crews that accompany patients on every AE flight, additional specialists comprising CCAT teams are involved in the evacuation of patients who need additional monitoring and care while in flight. Calculating the number of CCAT crews would resemble calculating the number of AE crews, except that CCAT crews would only accompany critical patients. Currently, JMAT reports the number of ambulatory and litter patients at each facility. It does not, however, provide a readily accessible total of the number of critical patients—the patients who may require the presence of a CCAT team. JMAT should be modified to report, for each location, the number of critical patients who are awaiting transport, and, in conjunction with the AE crew factors just listed, it should determine the requirement for CCAT crews.

CHAPTER FIVE

Conclusions and Recommendations

In this report, we have presented a framework for planning the expeditionary medical system that uses the patient flow rate, or STEP rate, as the guiding concept and unit of measurement. For treatment facilities, the STEP rate would be expressed as the number of patients treated per unit of time. For evacuation assets, the STEP rate would be expressed as the number of patients evacuated per unit of time. The STEP rate can be made more granular by specifying patient conditions (e.g., the number of trauma patients requiring surgery per day, the number of critical care patients requiring evacuation). Despite differences in detail and interpretation, the STEP rate ultimately is a measure of the number of patients flowing through the system per unit of time. As such, it applies to treatment facilities and evacuation assets across the military services and across all levels of care. Using this common unit of measurement helps integrate elements of the medical system that must be planned in conjunction with each other.

The STEP rate has been designed to be consistent with the current medical CONOPS, which emphasizes the movement of patients into ever-increasing levels of care. Likening patient treatment and evacuation to the flow of fluid through a system illustrates the interdependencies between treatment and evacuation and between the different levels of care. The flow of patients is optimized when treatment and evacuation rates at the different levels are balanced. Shortfalls in one function's capabilities will cause delays in patient flow, which will in turn affect the resources necessary for other functions at other levels. Using the STEP rate as a common unit of measurement throughout the medical system would facilitate integrated planning across treatment and evacuation functions, across the echelons of care, and across the military services. This integrated approach would allow planners to better balance the allocation of resources between the functions and among different locations, preventing waste in the deployment of assets in some areas and shortages in others.

The primary planning tool currently used by the joint community for medical planning, JMAT, is generally compatible with the planning concept presented in this report. To enable planners to fully implement the integrated approach to planning that we have presented, some modifications to JMAT will be necessary. These fall into three main areas:

- **Extend JMAT to include the entire chain of care.** This is necessary because all the functions—for each service and at every level of care—must be planned in a coordinated way.
- **Enhance JMAT modeling of patient holding capacity.** This will enable planners to better examine trade-offs between evacuation resources and treatment resources.
- **Enhance JMAT modeling of evacuation.** This will allow planners to identify the evacuation resources necessary to keep up with patient flow.

Investments in these areas would greatly enhance the effectiveness of medical planning and, ultimately, enable commanders to deploy the right resources in the right areas. This would allow the whole system to be brought up to operational capability more quickly and would ensure that patient care and transport occur with a minimum of delay.

We wish to remind the reader that, as is the case with using any concept of medical planning, effectively using the STEP rate concept depends on using planning assumptions that are appropriate for the chosen scenario. One set of assumptions concerns geography, which affects the flying times between levels of care and thereby affects the number of aircraft and crews needed for AE as well as what treatment facilities may be required at stopover locations. Planning tools should therefore be revised to allow users to vary assumptions related to flying times between the levels of care.

The second set of assumptions concerns the rate, timing, and type of injuries and diseases that are expected in a chosen scenario. Different types of scenarios will produce a different distribution of patient conditions. Medical planning, regardless of methodology, is only as good as the estimates of casualty numbers (and injury types) that underlie the planning assumptions. Different injuries or diseases require different levels and types of resources and may require different evacuation timing (e.g., faster evacuation or, conversely, more recovery time before evacuation). Casualty forecasts should therefore be updated to include lessons learned from recent operations, and they should be tailored to the different types of operations and scenarios that the U.S. military may face in the future.

The Need for Updated Casualty Estimates That Reflect a Spectrum of Scenarios

Casualty forecasting was not a main focus of our study, but it plays an important role in resource estimation: If casualty forecasts are off, resource estimates will be off. At best, a mismatch between forecast needs and actual needs would result in the overdeployment of certain resources, which could be wasteful. At worst, however, the mismatch would lead to unnecessary suffering on the part of injured service members because of a lack of treatment or evacuation resources.

In this appendix, we examine the assumptions underlying current casualty forecasting tools and assess the extent to which they take into account the full spectrum of scenarios that U.S. forces may face in the future. We also present analyses that illustrate the effect of inaccurate casualty forecasts on estimates of the medical resources needed to sustain specific levels of patient care.

The uncertain nature of real-world operations will always lead to uncertainty in casualty forecasting and resource estimation. Nonetheless, planners would be aided by improvements to casualty forecasting tools. Current tools are based on a limited set of scenarios and therefore may not accurately forecast other likely scenarios. Because inaccurate forecasts can have a major effect on estimates of resource requirements, current forecasting tools must be updated to take into account data from current operations and assumptions about the full spectrum of operational scenarios that U.S. military forces may face in the future, such as those presented in the Defense Planning Scenarios.

Casualty Estimation

Within JMAT, the user enters a set of casualty rates that determine the workload for medical care. Using the casualty rates, the model generates a number of casualties for each day, for each population. The casualties are distributed among the patient condition types listed in the PCOF, in accordance with a set of assumed percentages. These percentages vary based on service type (i.e., branch of the military) and the theater of operations. However, users also have the ability to enter their own table of PCOF percentages to distribute casualties across patient conditions for special groups, such as civilians or enemy prisoners of war.

Casualty rates drive the medical treatment facility demand within the JMAT model, and they have a significant effect on the simulation results. Therefore, it is extremely important to enter accurate casualty rates. However, there are great uncertainties in the casualty rates and in the underlying scenario assumptions, which means that multiple excursions and sensitivity runs are needed to develop a range of probable medical requirements.

Estimating Battle Casualties

There are four types of battle casualties: killed in action, wounded in action (WIA), missing, and captured. There are three accepted methods within the Department of Defense (DoD) for developing battle casualty estimates:

- an automated method that involves combat simulation
- a semi-automated or fully automated method that employs an interactive casualty estimation tool
- a manual method that relies on expert judgment and graphs of historical rates.

Unfortunately, none of these methods is universally accepted by all the military services. The discussion that follows pertains specifically to ground force casualty estimation.

The method favored by the Army is the first in the list: simulating combat using a force-on-force combat model and then collecting battle casualty information for further processing by a medical planning model. In general, this approach uses a computer algorithm to attrit the number of weapons and amount of equipment in a military unit in response to enemy contact. Battle casualties per weapon/amount of equipment destroyed are calculated using factors derived from historical data. Although this approach is reproducible and is often performed as part of omnibus force-sizing studies, combat models for large-scale operations have three distinct disadvantages in many contexts: (1) they are expensive, in terms of time and effort, to populate with data; (2) they aggregate the individual actions of small ground units into larger entities for the purpose of computational efficiency, making it difficult to determine where and when battle casualties occur; and (3) historical weapon- or equipment-based loss rates are less useful as current and future weaponry and tactics diverge further and further from those used in the past.

The second method is to estimate battle casualties using a purpose-built, interactive casualty estimation tool. The Marine Corps Casualty Estimation (CASEST) Model is an action-level (squad-to-battalion) deterministic simulation for projecting both battle casualties and casualties caused by accident and disease. CASEST's calculation of the casualty stream is based on the number and type of forces participating in a combat action, the intensity of the combat, and the duration of the action. In addition to providing numbers of casualties, CASEST reports the types of injuries incurred according to DoD patient condition codes.[1] CASEST also usually requires a large degree of detail to populate the model for a given scenario.

There are two variants of the third accepted method. In the "intensity" approach, the user assigns casualty rates to forces based on the intensity of combat and on the types of forces involved. The percentage of active combat elements is used as the gauge of intensity—the more combat elements engaged, the higher the intensity and the battle casualty rate. A criticism of this approach is that it considers neither the significant daily variability observed in histori-

---

[1]   A patient condition code is a standardized coding scheme that contains a unique identifying three-digit number, an injury description, and a treatment protocol appropriate for each level of medical care.

cal data nor the sizes of the forces involved (or their operational posture). The "rate patterns" approach, developed by George Kuhn, explicitly considers these factors when assigning rates. Given a planning time line expressed in days, forces are organized in PAR groupings, and rates are applied to those forces after considering the scheme of maneuver and the expected character of the opposing sides' interaction (also known as its *operational form*).[2] A distinguishing characteristic of this approach is the recognition that daily rates for engaged forces display pulses of higher rates and pauses of lower rates that reflect the local and episodic nature of combat.

Planning Assumptions Do Not Reflect Differences in Casualty Rates and Types

Today, many of the planning assumptions used in casualty forecasting are based on Cold War scenarios and major combat operations (MCO), particularly ground-based force-on-force combat. For example, the Ground Forces Casualty Forecasting System model developed by NHRC predicts casualties from ground battles using historical data from conflicts in Okinawa, Korea, Vietnam, and the Falklands.[3] Current planning assumptions presume that a well-developed basing infrastructure exists in such locations as Germany and Japan. Although the potential for MCO still exists in the form of major regional conflicts, such operations are unlikely to be appropriate for predicting casualties in the other types of scenarios that the U.S. military will face. In future years, U.S. forces may be called upon to conduct a wide range of operations, from MCO and irregular warfare to peacekeeping and humanitarian assistance, as reflected in the Defense Planning Scenarios developed by DoD. Casualty rates and patient mixes will differ with each type of scenario.

For example, current counterinsurgency operations in Iraq and Afghanistan, which primarily involve enemy guerilla attacks with small-arms fire and improvised explosive devices, exhibit a lower overall intensity of combat and a much lower casualty rate than would be expected during MCO, which might involve combat between two large infantry forces supported by armor and artillery. Also, the timing of patient arrivals varies with different operational scenarios: In combat scenarios, casualties will be produced during the duration of the conflict, but, in disaster relief operations, the bulk of the casualties may be produced all at once, during the disaster incident itself. (For more information about injuries sustained during disaster relief operations, see Appendix B.)

Different operational scenarios lead to different types of casualties and to different casualty rates. For instance, compared with combat operations, humanitarian assistance operations can result in a much higher proportion of patients with diseases (as opposed to traumatic injuries). These diseases are often associated with existing chronic conditions rather than with the disaster itself. Even in MCO scenarios, changes in weapon technology, in the protective equipment worn by service members, and in the medical care available have an effect on the mix of patient conditions seen: Injuries occur in different areas of the body than in previous conflicts, and patients who might have died on the battlefield in an earlier war now make it

---

[2] For an in-depth discussion of the rate patterns approach, see G. W. S. Kuhn, *CJCS Guide to Battle Casualty Rate Patterns for Conventional Ground Forces*, The Joint Staff, January 15, 1998.

[3] See C. G. Blood, J. M. Zouris, and D. Rotblatt, *Using the Ground Force Casualty Forecasting System (FORECAS) to Project Casualty Sustainment*, Naval Health Research Center, Report No. 97-39, 1997. Note that NHRC is updating its modeling tools to reflect more-recent data from the operations in Iraq and Afghanistan.

to medical facilities for treatment.[4] Consequently, the medical capabilities required for current and future MCO will likely differ from the medical capabilities formerly required for Cold War–era operations.

Differences in casualty rates and the mix of patient types pose different challenges in terms of the medical resources that will be required. It may appear at first that planning based on MCO assumptions would be a safe approach to setting resource levels for medical support, since all other operations could be considered lesser included cases. However, resourcing to the higher rate or mix of casualties associated with MCO might lead to an overbuilding of medical capability, and overbuilding comes at a cost; for example, higher casualty assumptions would require larger facilities, which could be slower to deploy and consume valuable airlift resources, particularly early in an operation.

Figure A.1 illustrates the limitations of using historical MCO data as a predictor of casualties across the full spectrum of operations. We derived the frequency of patient conditions for MCO scenarios using historical data contained in JMAT. We then compared the frequency of these conditions during MCO with their frequency during a counterinsurgency operation (specifically, Phase 1 of Operation Iraqi Freedom) and during humanitarian assistance operations.

The forecasts in Figure A.1 indicate that the mix of patient conditions will vary with the type of scenario; for example, more diseases occur under the humanitarian assistance operations scenario than under the MCO scenario. A forecast based on historical MCO data would not capture such differences and would lead to faulty estimates of the resources needed to treat and evacuate patients. Because the distribution of patient conditions varies with the operational scenario, the medical resources that are needed will also vary with the operational scenario.[5]

The Sensitivity of Resource Estimates to Patient Conditions

We evaluated the sensitivity of medical resource requirements to the distribution of patient conditions (i.e., the frequency with which head injuries, burns, diseases, or psychiatric conditions occur) under the three scenarios previously discussed: an MCO (based on assumptions built into JMAT), a counterinsurgency operation (based on casualty data compiled by NHRC using the Expeditionary Medical Encounter Database), and a notional humanitarian assistance scenario derived from an amalgamation of humanitarian assistance operations that we compiled.[6] We computed the number of hours that 100 patients from each scenario spend in treatment areas (ERs, ORs, etc.) at a Level 3 medical facility. Figure A.2 presents the results of this analysis.

Our analysis indicates that ER and OR hours are not sensitive to patient conditions; in other words, patients in all three scenarios spend about the same amount of time in these areas. However, the hours spent in intensive care units, intermediate care wards, and minimal care wards are more sensitive to the distribution of patient conditions. For example, the

---

[4]   See, for instance, B. Owens, J. F. Kragh, J. C. Wenke, J. Macaitis, C. Wade, and J. B. Holcomb, "Combat Wounds in Operation Iraqi Freedom and Operation Enduring Freedom," *The Journal of Trauma-Injury, Infection, and Critical Care,* Vol. 64, No. 2, February 2008, pp. 295–299.

[5]   Note, however, that some patient conditions may arise regardless of whether they resulted from the patient being wounded in action or sustaining a nonbattle injury. Thus, some patient conditions would be seen in both scenarios, albeit for different reasons.

[6]   We generated this third scenario to illustrate how a humanitarian assistance scenario could result in different injuries than those seen in a wartime scenario. However, our notional scenario is not intended to be interpreted as a prediction of a particular relief operation. See Appendix B for more information about the methodology used to generate this example.

Figure A.1
Variations in Casualty Forecasts for Three Operational Scenarios

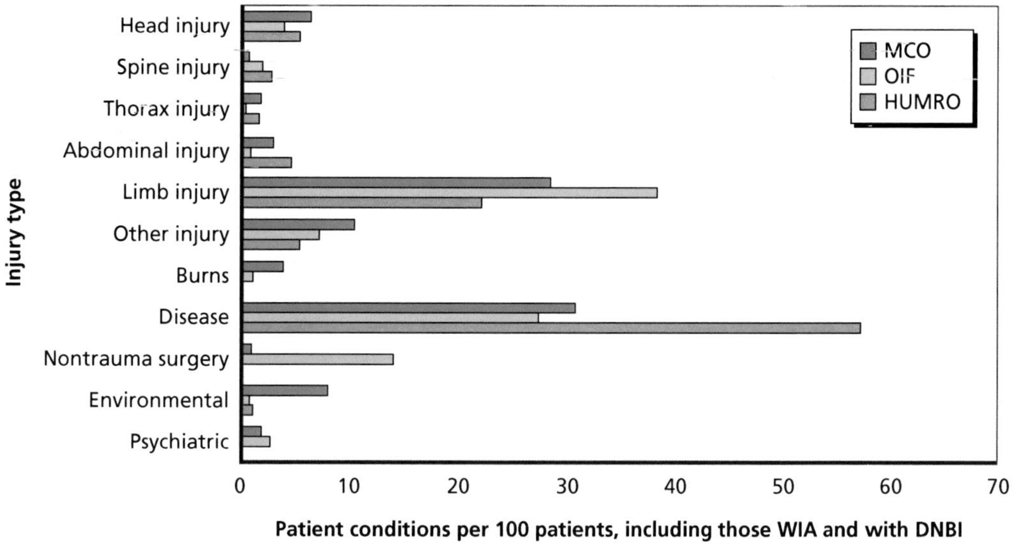

SOURCE: Frequencies for Operation Iraqi Freedom are derived from our analysis of historical data produced by NHRC. Frequencies for the humanitarian relief operations scenario represent an amalgamation of data drawn from several recent disaster relief operations (see Appendix B for details).

NOTES: We have normalized the injury distribution to 100 patients for each scenario. HUMRO = humanitarian relief operations. OIF = Operation Iraqi Freedom.

RAND TR1003-A.1

100 patients in the counterinsurgency scenario (i.e., Phase 1 of Operation Iraqi Freedom) spend about 2,220 hours in a minimal care ward, whereas the 100 patients under the MCO scenario spend about 1,600 hours there, a difference of approximately 600 hours. This analysis suggests that the medical resource requirements associated with minimal care wards will be greater for patients in counterinsurgency operations than for patients in an MCO scenario.

In summary, this analysis highlights both the sensitivity of results to the parameters used in casualty forecasting and the importance of considering the range of factors that influence the medical requirements needed in theater.

Conclusion

Planning tools, such as JMAT, contain assumptions about the mix of patient types. The user inputs the population size and the overall daily injury rates, but the breakdown of those injuries into various patient conditions and levels of severity is based on proportions that are built into the software. It is often not clear in the software or its documentation whence these proportions are derived, and interviews with model developers indicate that the injury mix assumptions are based on potentially outdated historical data from as far back as World War II. At the very least, planners need to be aware of the assumptions built into casualty forecasting tools and medical planning tools. Ideally, however, they will be given an opportunity to adjust these patient mix percentages using updated information appropriate to the scenario being modeled.

Figure A.2
Sensitivity of Medical Resources to Operational Scenarios

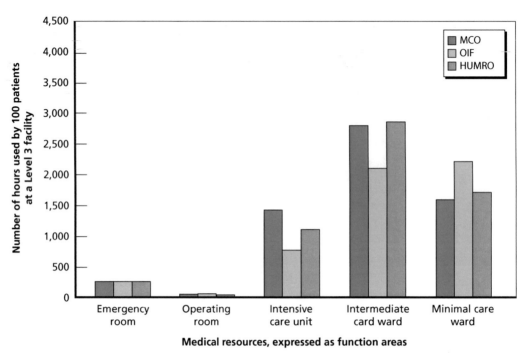

RAND *TR1003-A.2*

Modeling Medical Requirements for Humanitarian Assistance Operations

Effective planning of medical requirements requires forecasts of casualty rates, injury types, and injury severity that are appropriate to the scenario being planned. The injuries seen in humanitarian assistance operations can be quite different from those seen in combat operations. However, casualty forecasting models are generally based on historical data from major combat operations.

In the course of our study, we constructed an example patient distribution for a humanitarian assistance scenario. It is not intended to serve as an actual model or forecast of casualties for all future humanitarian assistance operations. Rather, it was built to illustrate the potential differences in the types of injuries seen and, therefore, in the types of medical resources that would be required. In this appendix, we explain how we constructed this sample scenario and present it as an example of how a model for forecasting humanitarian assistance medical requirements may be built.[1]

Differences Between Humanitarian Assistance and Combat Operations That Influence Medical Support Requirements

Several differences between humanitarian assistance and combat operations lead to important distinctions in medical assistance requirements. Key among these are differences in the affected populations, the predictability and timing of incidents, and the types of hazards and resulting injury patterns seen.

Differences in Affected Populations
The populations affected by humanitarian disasters are very different from those participating in U.S. military combat operations. Humanitarian disasters generally affect the entire population of the areas they strike, whereas the U.S. military service members constitute a sample that is more homogeneous and healthier than the general population. Populations affected by humanitarian disasters have a wider age range, greater incidence of preexisting medical conditions, and lower levels of fitness and overall health. These differences result in a wider range of necessary treatment regimes, differences in the types of medical supplies and

---

[1] NHRC, in response to tasking from the Office of the Secretary of Defense, is currently developing a PCOF for humanitarian assistance operations.

equipment needed, and greater variability in patient outcomes and complications compared with those that the military medical system is optimized to handle.

One such challenge is the greater vulnerability of the very young and very old to injury and death in disasters.[2] Military medical operations involving pediatric cases in particular are difficult and resource intensive.[3] Another example raised during interviews with Air Force medical personnel who assisted in the response to the 2004 Asian tsunami is that a high demand for insulin led to shortages and to urgent calls for more supplies. It is important to note that the greater vulnerability of the young and old in disasters is not restricted to disasters in poor countries: In the two weeks following Hurricane Katrina, 56 percent of evacuees residing in shelters had arrived at the shelter with a chronic disease requiring medication.[4]

Differences in the Predictability and Timing of Incidents

Although the pace and intensity of military combat operations vary considerably, casualties arrive in a semicontinuous fashion. In addition, military combat is generally preceded by planning and deployment of medical resources. Humanitarian disasters, in contrast, occur with little or no warning and are characterized by a short pulse with a high casualty rate. This means that a large flux of casualties present to a potentially degraded or overwhelmed medical system immediately following an incident. Outside assistance is usually at least hours—and often days—away, meaning that the medical system is weakest when demand is highest. Military medical assistance in these situations may therefore be very different from that available in military combat operations. Experience has shown that, compared with combat situations, disaster situations entail far less emphasis in military medical assistance on triage, trauma care, emergency surgery, and other forms of urgent care. This is because the demand for such capabilities will already have expired by the time the military medical assistance arrives.[5]

Differences in Types of Hazards and Resulting Injury Patterns

A third difference between natural disasters and combat operations is the causes of injury and the resulting distributions of injury types. Military injuries tend to be dominated by wounds from firearms and explosives, whereas injuries from natural disasters are caused by the very different hazards associated with earthquakes, hurricanes, tsunamis, and other disasters. Such differences are expected to lead to different injury patterns, although the extent of such differences is uncertain because injury distributions from natural disasters are not well characterized. Some experience suggests that there may be less demand for surgery in natural

---

[2] See Y. Bar-Dayan, A. Leiba, P. Beard, D. Mankuta, D. Engelhart, et al., "A Multidisciplinary Field Hospital as a Substitute for Medical Hospital Care in the Aftermath of an Earthquake: The Experience of the Israeli Defense Forces Field Hospital in Duzce, Turkey, 1999," *Prehospital and Disaster Medicine*, Vol. 20, No. 2, 2005, pp. 103–106; S. Doocy, C. Robinson, C. Moodie, and G. Burnham, "Tsunami-Related Injury in Aceh Province, Indonesia," *Global Public Health*, Vol. 4, No. 2, March 2009, pp. 205–214; and N. J. Liang, Y. T. Shih, F. Y. Shih, H. M. Wu, H. J. Wang, et al., "Disaster Epidemiology and Medical Response in the Chi-Chi Earthquake in Taiwan," *Annals of Emergency Medicine*, Vol. 38, 2001, pp. 549–555.

[3] P. C. Spinella, M. A. Borgman, and K. S. Azarow, "Pediatric Trauma in an Austere Combat Environment," *Critical Care Medicine*, Vol. 36, Supplement 7, July 2008, pp. S293–S296.

[4] P. G. Greenough, M. D. Lappi, E. B. Hsu, S. Fink, Y. H. His, et al., "Burden of Disease and Health Status Among Hurricane Katrina–Displaced Persons in Shelters: A Population-Based Cluster Sample," *Annals of Emergency Medicine*, Vol. 51, No. 4, 2008, pp. 426–432.

[5] T. F. Haley and R. A. De Lorenzo, "Military Medical Assistance Following Natural Disasters: Refining the Rapid Response," *Prehospital and Disaster Medicine*, Vol. 24, No. 1, 2009, pp. 9–10.

disasters than in combat operations.[6] As discussed in more detail later in this appendix, injuries in natural disasters appear to be dominated by lacerations and infected wounds, respiratory infections, gastrointestinal illnesses, complications from unattended chronic conditions, and non–disaster-related injuries.

Casualty Distributions in Natural Disasters

To help understand capability needs for military medical assistance in humanitarian disasters, planners need to know the medical treatment requirements for deployed units. These requirements are influenced by several factors, such as the type of disaster and resulting distribution of injuries, the arrival time of the unit, and the military's role relative to local and other deployed medical resources. This subsection addresses the distribution of injury types in disasters. Logistical factors, such as deployment times and the military's role in humanitarian medical assistance, are equally important, but they are both complex and the subject of ongoing exploration that goes beyond the scope of this analysis.

We conducted a literature search for studies presenting information about injury types generated by natural disasters. We searched the PubMed and Web of Knowledge databases using combinations of the following terms: *disaster, earthquake, hurricane, tsunami, flood, injury, casualty,* and *epidemiology.* The search covered papers published from 1999 through 2009. Few of the studies we identified report quantitative estimates of injury or illness distributions in disasters. Most of the estimates we did find concerned, in descending order, earthquakes, hurricanes, and the 2005 Indonesian tsunami. Most studies we identified used a retrospective review of medical records, but other methods (including real-time review of medical records, surveillance reporting, and surveys) were also used. The samples and time spans covered by the studies vary considerably. Samples ranged from single medical clinics to multistate data systems; time spans ranged from days to months, starting as early as the incident outset and as late as several months after the incident.

Although we found a relatively large number of studies addressing the epidemiology of disaster-related injuries and illnesses, most focus on demographic information and other associations among the injured. Surprisingly few studies report information about the injuries or illnesses themselves. Thus, there are few data available that can be used to estimate the total distributions of injury types in disasters. Because there are so few data, we chose to combine data from different disaster types in our analysis.

Injuries

Lacerations and fractures appear to be the most common injury types seen in natural disasters.[7] Crush-related injuries can also be significant in earthquakes.[8] Many of the lacerations and other

---

[6] Bar-Dayan et al., 2005.

[7] K. Ashraf Ganjouei, L. Ekhlaspour, E. Iranmanesh, P. Poorian, S. Sohbati, N. Ashraf Ganjooei, F. Rashid-Farokhi, and S. Karamuzian, "The Pattern of Injuries Among the Victims of the Bam Earthquake," *Iranian Journal of Public Health,* Vol. 37, No. 3, 2008, pp. 70–76; Doocy et al., 2009; K. M. McNeill, P. Byers, T. Kittle, S. Hand, J. Parham, et al., "Surveillance for Illness and Injury After Hurricane Katrina—Three Counties, Mississippi, September 5–October 11, 2005," *MMWR Weekly,* Vol. 55, No. 9, March 10, 2006, pp. 231–234; J. M. Mulvey, S. U. Awan, A. A. Qadri, and M. A. Maqsood, "Profile of Injuries Arising from the 2005 Kashmir Earthquake: The First 72 h.," *International Journal of the Care of the Injured,* Vol. 39, 2008, pp. 554–560; T. Prasartritha, R. Tungsiripat, and P. Warachit, "The Revisit of 2004 Tsunami in Thailand: Characteristics of Wounds," *International Wound Journal,* Vol. 5, 2008, pp. 8–19.

[8] M. Bulut, R. Fedakar, S. Akkose, H. Ozguc, and R. Tokyay, "Medical Experience of a University Hospital in Turkey After the 1999 Marmara Earthquake," *Emergency Medicine Journal,* Vol. 22, 2005, pp. 494–498.

wounds become infected, including with tetanus.[9] The high incidence of infected wounds may reflect delays in access to appropriate medical attention.

Although several studies report the occurrence frequency of one or more of the most common injury types, few studies present complete distributions. Findings from three studies that provide moderately comprehensive distributions of injury types from different types of disasters are presented in Figure B.1. These results are consistent with the majority of the studies we reviewed in that they report that lacerations and fractures are the most common injury type.

Another way to characterize injuries is in terms of the body part injured. This perspective complements injury type in that it can provide additional insights about injury severity (e.g., injuries to extremities are often less severe than injuries to the abdomen or chest); however, it says nothing about the type of injury. Reasonably comprehensive data for four earthquakes are presented in Figure B.2. (We found no data for other types of disasters.) The results in Figure B.2 indicate that earthquakes most often cause lower-limb injuries, then upper-limb and head injuries.

Figure B.1
Reported Distribution of Injury Types Resulting from Three Natural Disasters

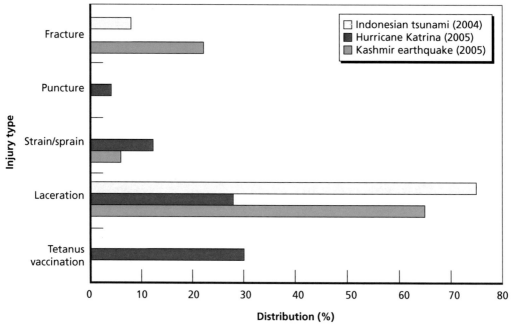

SOURCES: Doocy et al., 2009; McNeill et al., 2006; Mulvey et al., 2008.
RAND *TR1003-B.1*

---

[9]   Bulut et al., 2005; S. W. Fan, "Clinical Cases Seen in Tsunami Hit Banda Aceh—From a Primary Health Care Perspective," *Annals, Academy of Medicine*, Singapore, Vol. 35, 2006, pp. 54–59; McNeill et al., 2006; Mulvey et al., 2008; L. J. Redwood-Campbell and L. Riddez, "Post-Tsunami Medical Care: Health Problems Encountered in the International Committee of the Red Cross Hospital in Banda Aceh, Indonesia," *Prehospital and Disaster Medicine*, Vol. 21, No. 1, 2006, pp. 1–7; R. A. Streuli, "Tsunami in South-East Asia—Rapid Response Deployment in Banda Aceh," *Ther Umsch*, Vol. 65, No. 1, January 2008, pp. 15–21.

Figure B.2
Reported Distribution of Injured Body Parts Resulting from Four Earthquakes

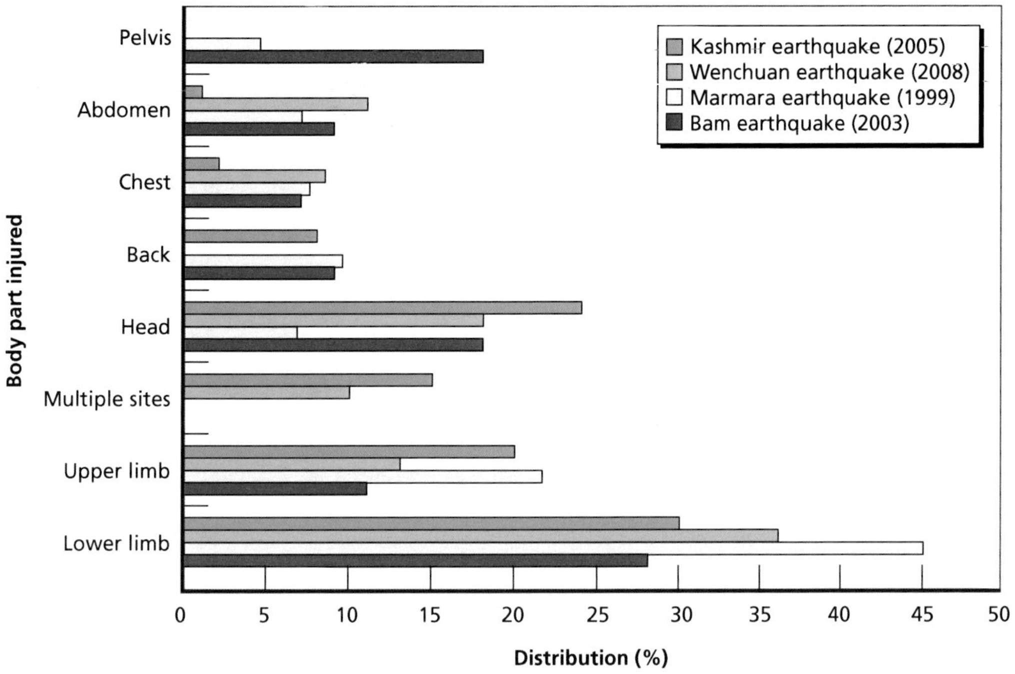

SOURCES: Ashraf Ganjouei et al., 2008; Bulut et al., 2005; Mulvey et al., 2008; Lei Zhang, He Li, Janis R. Carlton, and Robert Ursano, "The Injury Profile After the 2008 Earthquakes in China," *Injury*, Vol. 40, 2009, pp. 84–86.
RAND *TR1003-B.2*

Illnesses

Studies of disaster epidemiology generally tend to address injuries. In our review, we found that studies that examine illnesses often provide descriptions of types of illnesses, possible causes, and treatments, but they rarely provide more than qualitative statements about the proportions of different types of illnesses. Further, the available information reflects widely variable situations in different disasters. Hence, there are few estimates—or even qualitative generalities—available. One review of the epidemiology of hurricanes found that, although outbreaks of infectious diseases have occurred after hurricanes, such outbreaks are not particularly common and do not constitute a major risk.[10] Studies of Hurricane Katrina suggest that more important than treating infectious disease outbreaks is caring for patients with preexisting medical conditions and providing them with refills of their prescription medication.[11] Several studies of the 2004 Asian tsunami and of Hurricane Katrina note the common occurrence of gastrointestinal ailments, respiratory infections, skin infections and rashes, and infected wounds.[12] We found virtually no information on illnesses following earthquakes.

---

[10] J. M. Schultz, J. Russell, and Z. Espinei, "Epidemiology of Tropical Cyclones: The Dynamics of Disaster, Disease, and Development," *Epidemiologic Reviews*, Vol. 27, 2005, pp. 21–35.

[11] Greenough et al., 2008; McNeill et al., 2006.

[12] R. J. Brennan, and K. Rimba, "Rapid Health Assessment in Aceh Jaya District, Indonesia, Following the December 26 Tsunami," *Emergency Medicine Australasia*, Vol. 17, No. 4, August 2005, pp. 341–350; Doocy et al., 2009; Fan, 2006; McNeill et al., 2006; Redwood-Campbell and Riddez, 2006; F. Riccardo, L. E. Pacifici, A. G. De Rosa, E. Scaroni,

Relative Proportions of Injuries and Illnesses

Another poorly documented statistic is the relative proportions of injuries and illnesses resulting from natural disasters. We found only three studies that report relative proportions of injuries and illnesses, and two of those studies treat the same incident. These results, presented in Figure B.3, show a wide variation in injury and illness proportions. The average of the three studies is 40 percent injuries and 60 percent illnesses.

Modeling Patient Conditions for Humanitarian Disasters

To help planners understand the potential military medical requirements for humanitarian assistance operations and how those differ from requirements for military combat operations, we used findings from our literature review to estimate a military patient condition occurrence frequency distribution for humanitarian medical assistance. Given limitations in the amount and reliability of availability data, and given the great variability in injury distributions in disasters, any estimated distribution—including ours—must be viewed as highly uncertain. Of course, similar uncertainties exist in the case of military combat operations, and estimated patient condition occurrence frequency distributions should always be used only as general

Figure B.3
Reported Percentages of Injuries and of Illnesses and Diseases Resulting from Two Disasters

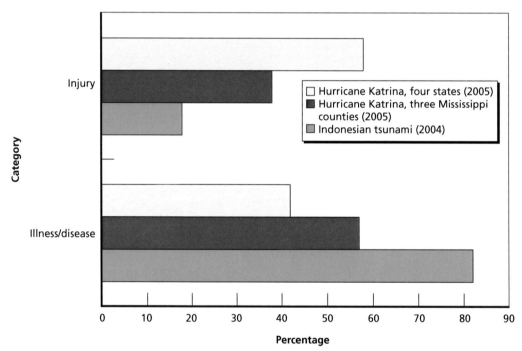

SOURCE: F. Averhoff, S. Young, J. Mott, A. Fleischauer, J. Brady, et al., "Morbidity Surveillance After Hurricane Katrina: Arkansas, Louisiana, Mississippi, and Texas, September 2005," *MMWR Weekly*, Vol. 55, No. 26, July 7, 2006, pp. 727–731; Y. H. Kwak, S. Do Shin, K. S. Kim, W. Y. Kwon, and G. J. Suh, "Experience of a Korean Disaster Medical Assistance Team in Sri Lanka After the South Asia Tsunami," *Journal of Korean Medical Science*, Vol. 21, No. 1, 2006, pp. 143–150; McNeill et al., 2006.
RAND *TR1003-B.3*

L. Nardi, et al., *Epidemiological Surveillance: A Growing Role in Humanitarian Emergencies*, 5th International Congress on Tropical Medicine and International Health, Amsterdam, May 24–28, 2007; Streuli, 2008.

guidance. Indeed, planners are aware that they must be prepared for specific conditions that differ greatly from planning factors. Nonetheless, comparing estimates for humanitarian and military medical operations may provide useful general insights about differences in medical requirements.

We treated injuries and illnesses separately. For injuries, we linked the injury data from the literature to patient condition codes by body part. The distribution of injuries for each body part in Figure B.2 was renormalized to exclude multiple-site injuries, and the resulting distribution is shown in Table B.1. The percentage for each body part was distributed among all the patient condition codes associated with that body part.[13] The resulting occurrence frequency for each patient condition code was then weighted by its priority rating. (Each patient condition code is assigned a numerical priority rating from one to four, with one having the highest priority for treatment and four having the lowest priority. Hence, priority rating is inversely proportional to injury or illness severity.) We obtained the patient condition codes and associated priority weights from data files built into JMAT. Weighting by priority rating thus gives higher weight to the less-severe injuries. We also used this weighting to approximate the greater frequency of less-severe injuries relative to more-severe injuries.

In the case of illnesses, given the lack of relevant quantitative data, we simply distributed illnesses evenly among the four most common illness types identified in the literature. The results are shown in Table B.2. Each of the four illness classifications is associated with a cluster of patient conditions. As we did for injuries, we generated occurrence frequencies for the illness patient condition codes by dividing the percentages in Table B.2 by the number of patient conditions associated with that illness. We then weighted the resulting occurrence frequency of each patient condition code by its priority rating.

As shown in Figure B.3, the relative proportions of injuries and illnesses were 40 percent and 60 percent, respectively. The resulting occurrence frequency distribution for the 167 injury

Table B.1
Distribution of Injured Body Parts (Excluding Multiple-Site Injuries) in Four Earthquakes

Body Part	Occurrence Frequency
Lower limb	35%
Upper limb	18%
Head	16%
Back	9%
Chest	6%
Abdomen	6%
Pelvis	11%
Total	100%

NOTE: Due to rounding, percentages do not add to exactly 100.

---

[13] For example, head injuries account for 7 percent of all injuries, and there are 23 different patient condition codes for various head injuries. Each head injury patient condition code therefore receives an initial, unweighted occurrence frequency of 7% ÷ 23 = 0.3%.

Table B.2
Estimated Distribution of Illness Types in Natural Disasters

Illness Type	Occurrence Frequency
Gastrointestinal	25%
Respiratory	25%
Skin infection/rash	25%
Infective/parasitic	25%
Total	100%

and illness patient conditions observed in our data from natural disasters is summarized in Table B.3. Patient conditions in the table are aggregated in clusters by affected body part (for injuries) or by disease group (for illnesses). The full occurrence frequency distribution of individual patient conditions is available from the authors upon request.

Timing of Injuries

Our literature search also revealed some information about the timing of injuries in natural disasters. As noted earlier, most injuries in natural disasters present in the first several days following an incident. Few of the studies we identified sampled a time period that included the first day of the disaster and went on long enough to allow for calculation of an estimate of the total arrival distribution. We found three studies, all of earthquakes, that provide a relatively complete picture of the distribution of arrivals of injured victims. These results are shown in Figure B.4.

These results show that injured patients arrive at medical facilities very shortly after the earthquake. The earliest that U.S. military medical assistance can be anticipated to arrive and be operational in response to a disaster is approximately 48–96 hours after the earthquake.[14] By that time, half or more of the injuries will have been seen in existing local medical facilities. Injury arrivals due to other disaster types, such as hurricanes, tsunamis, and floods, may be spread over longer periods.[15] The extent to which later arrivals reflect injuries actually occurring at later times (rather than delays between receiving an injury and seeking treatment) is unclear.

Conclusion

Injury type, severity, and frequency will vary depending on the scenario. These differences have implications for the medical support that will be required. Consequently, it is vital for medical planners to have casualty forecasts appropriate to the scenario being planned.

In this appendix, we have shown how we created an example forecast of patient types for a humanitarian assistance scenario. The example scenario represents an amalgam of different

---

[14] R. Malish, D. E. Oliver, R. M. Rush, E. Zarzabal, M. J. Sigmon, et al., "Potential Roles of Military-Specific Response to Natural Disasters—Analysis of the Rapid Deployment of a Mobile Surgical Team to the 2007 Peruvian Earthquake," *Prehospital and Disaster Medicine*, Vol. 24, No. 1, 2009, pp. 3–8; U.S. Government Accountability Office, *Homeland Defense: Planning, Resourcing, and Training Issues Challenge DOD's Response to Domestic Chemical, Biological, Radiological, Nuclear, and High-Yield Explosive Incidents*, Washington, D.C., GAO-10-123, October 2009.

[15] See, for example, McNeill et al., 2006.

Table B.3
Summary of Modeled Distribution of Patient Conditions for Injuries and Illnesses in Natural Disasters

Body Part or Disease Group	Occurrence Frequency
Head	0.0527
Back	0.0286
Upper extremities	0.0783
Thorax	0.0189
Abdomen	0.0148
Pelvis	0.0319
Lower extremities	0.1078
Dermatological	0.1761
Respiratory	0.1725
Gastrointestinal	0.1702
Infective/parasitic	0.1482
Sum	1.000

Figure B.4
Reported Distribution of Injury Arrival Times in Three Large Earthquakes

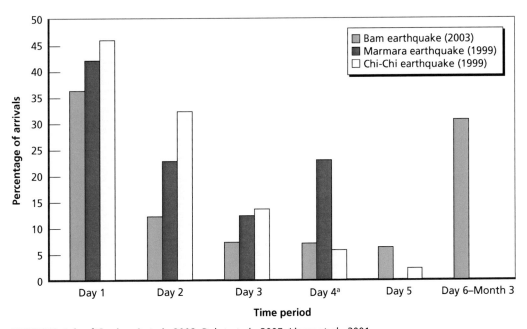

SOURCES: Ashraf Ganjouei et al., 2008; Bulut et al., 2005; Liang et al., 2001.
[a]In the case of the Marmara earthquake, the period is Day 4–Week 7.
RAND TR1003-B.4

types of disasters, however, since there were insufficient data on each type of disaster to permit us to model each type separately. The resulting scenario is intended to illustrate how humanitarian assistance may differ from wartime operations and should not be interpreted, per se, as a predictor of future requirements for humanitarian assistance. Nonetheless, the process we used in building this example may serve as a guide to constructing better casualty forecasts for these types of assistance operations.

Considerations in Selecting Stopover Points for Aeromedical Evacuation

Future operations could require the Air Force Medical Service to provide medical support in areas located far from existing medical infrastructure. Although mobile facilities would be deployed in theater, the extended flying times associated with evacuating patients from these forward facilities to more-definitive care could require adding stopover points to the evacuation route. In this appendix, we describe some of the issues that would arise in planning evacuation with stopovers. Our discussion uses AE in Africa to illustrate some of these issues.

One approach to selecting AE stopover points is to consider the level of medical capabilities available. Interviews with Air Force aeromedical evacuation specialists indicate that patients would never be moved to a lower level of care. Patients from Iraq and Afghanistan are evacuated out of an in-theater Level 3 facility; under such conditions, a stopover location would need to be Level 3 or higher.

There are different ways to accommodate the need for medical capabilities. One is to stop over in a location with a U.S. military medical facility. However, such facilities are, understandably, concentrated in areas where U.S. military personnel are stationed, and they tend not to be located in some of the remote areas that would be needed for an AE flight stopover. Review of the TRICARE website shows that the U.S. military medical facilities closest to the African continent are in Spain, Italy, Turkey, and Bahrain.

Medical capabilities at stopover points could also be provided by host country medical facilities. However, we have not been able to identify a database of medical facilities that could be used to select sites appropriate for AE stopovers. Airfield site surveys from U.S. Transportation Command contain very little information about associated medical facilities. Information logged into the U.S. Transportation Command Regulating and Command and Control Evacuation System (TRAC2ES) has generated an extensive AE database that includes information about medical treatment facilities. However, a major shortcoming of the TRAC2ES database is that it is built using operational data and, thus, includes information only about sites the U.S. military has used during prior aeromedical evacuations: It therefore provides little help in identifying and characterizing potential new locations. One potential avenue for identifying appropriate host nation medical treatment facilities is through medical contractors. TRICARE contracts with International SOS (a private company that arranges overseas medical care and evacuation for its clients) to provide health care services for U.S. military personnel around the world, including several locations in Africa. U.S. Africa Command, the combatant command with responsibility for Africa, also contracts with International SOS for peacetime AE services in Africa. Trusted contractors with detailed local knowledge may be good sources of information used to characterize and select medical facilities appropriate for AE stopover points. One complication associated with using host nation medical facilities is obtaining permission from the host countries and arranging working relationships with the local facilities. Such issues may take time and may negatively affect patient care or throughput rates.

A third source of medical resources at AE stopover points is deployable U.S. military medical facilities, such as Army Combat Support Hospitals, Air Force EMEDS, or Navy expeditionary medical facilities. This approach's advantage is that these military assets can be deployed almost anywhere, so AE stopover points could be colocated with strategic airlift hubs. Alternatively, deployed medical resources could be set up adjacent to host country medical facilities to augment their capacity or capability. Employing host country medical facilities in this manner would still require addressing the diplomatic and operational issues just noted. In addition, deployed medical resources are primarily designed for use in theater, and allocating additional units to AE stopover points would put additional strain on a limited resource.

Another approach for selecting AE stopover points is to colocate them with hubs used for strategic airlift—airfields where large aircraft coming from the United States or Europe can land and transfer their cargo or passengers to smaller aircraft for further transport to their final destination. Colocating medical facilities with these hubs is desirable because it puts patients on the same routes as other cargo and passengers, thereby minimizing the need for aircraft to make special flights just to transport patients. However, no strategic airlift plans have been developed for operations to and from or within Africa,[1] so no such hubs have been agreed upon. Current planning assumes that the only permanent airlift infrastructure on the African continent will be located at Camp Lemonier, Djibouti. The TRICARE contract with International SOS does not currently cover Djibouti; the closest covered site is Addis Ababa, Ethiopia, which is 350 miles from Djibouti. According to TRAC2ES, Djibouti has a Level 2 French military hospital, the Bouffard Army Hospital Center, and, as of January 2009, a Level 3 U.S. Navy ten-bed Expeditionary Medical Facility was located in the country. There is no information about the Bouffard Army Hospital Center in TRAC2ES or in any of several hospital databases available on the Internet.

Although Djibouti is the only current or planned airlift hub in Africa, it is very well located for acting as an AE stopover point. As illustrated in Figure C.1, Djibouti is within 2,800 nautical miles of nearly all of Africa and is 2,870 nautical miles from Landstuhl, Germany. Military personnel wounded almost anywhere in Africa could be evacuated through Djibouti to Landstuhl. Such a system would, however, depend on having sufficient medical capabilities in Djibouti. Depending on the level of existing capabilities and the capabilities needed for anticipated operations, medical resources in Djibouti may need to be augmented to support an AE stopover point.

Air Mobility Command's "Global En Route Strategy" explores provisional strategic airlift hubs for South America and identifies Palanquero, Colombia, as a cooperative security location.[2] Palanquero is within 2,800 nautical miles of nearly all of South America and the eastern United States (see Figure C.2). In terms of location, Palanquero is therefore well positioned as an AE stopover point for people injured in South America. In terms of medical facilities, however, it is not ideal. The closest U.S. military medical facilities are in Cuba and Puerto Rico. The only Colombian medical facilities listed in TRAC2ES are in Arauca and Cartagena, both

---

[1]   Air Mobility Command, "Global En Route Strategy," white paper, 2009.

[2]   Air Mobility Command, 2009.

Figure C.1
The 2,800–Nautical Mile Radius Around Djibouti

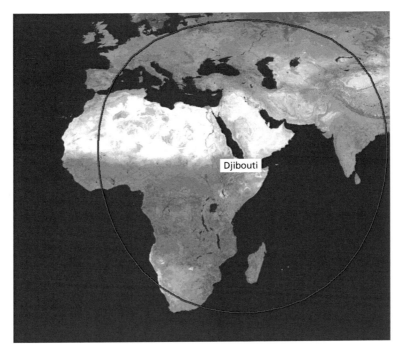

RAND *TR1003-C.1*

Figure C.2
The 2,800–Nautical Mile Radius Around Palanquero, Colombia

RAND *TR1003-C.2*

more than 250 miles from Palanquero. Bogotá is about 75 miles away. If Palanquero were to be used as an AE stopover point, medical resources may need to be deployed to provide the needed capability.

As the Air Force considers locations around the world for use as airlift hubs, the Air Force Medical Service should consider both how medical support can be provided to these hubs and how these hubs may enable medical support to be extended to regions of the world far from existing U.S. military medical infrastructure. Anticipated AE needs, estimated through the use of casualty forecasting models that properly reflect specific scenarios, will help determine how important access to medical capabilities are in selecting hub sites. Selecting AE stopover points in conjunction with strategic airlift hubs would help constrain the number of location options that must be investigated.

Bibliography

Air Mobility Command, "Global En Route Strategy," white paper, 2009.

Akimeka, *Joint Medical Analysis Tool (JMAT) Version 1.0.1.0 Software Users Manual*, Version 1.7, December 20, 2007.

American College of Surgeons Committee on Trauma, *Resources for Optimal Care of the Injured Patient 2006*, 2006.

Ashraf Ganjouei, K., L. Ekhlaspour, E. Iranmanesh, P. Poorian, S. Sohbati, N. Ashraf Ganjooei, F. Rashid-Farokhi, and S. Karamuzian, "The Pattern of Injuries Among the Victims of the Bam Earthquake," *Iranian Journal of Public Health*, Vol. 37, No. 3, 2008, pp. 70–76.

Averhoff, F., S. Young, J. Mott, A. Fleischauer, J. Brady, et al., "Morbidity Surveillance After Hurricane Katrina: Arkansas, Louisiana, Mississippi, and Texas, September 2005," *MMWR Weekly*, Vol. 55, No. 26, July 7, 2006, pp. 727–731.

Bar-Dayan, Y., A. Leiba, P. Beard, D. Mankuta, D. Engelhart, et al., "A Multidisciplinary Field Hospital as a Substitute for Medical Hospital Care in the Aftermath of an Earthquake: The Experience of the Israeli Defense Forces Field Hospital in Duzce, Turkey, 1999," *Prehospital and Disaster Medicine*, Vol. 20, No. 2, 2005, pp. 103–106.

Blood, C. G., and E. D. Gauker, *The Relationship Between Battle Intensity and Disease Rates Among Marine Corps Infantry Units*, Naval Health Research Center, Report No. 92-1, 1992.

Blood, C. G., D. K. Griffith, and C. B. Nirona, *Medical Resource Allocation: Injury and Disease Incidence Among Marines in Vietnam*, Naval Health Research Center, Report No. 89-36, 1989.

Blood, C. G., J. M. Zouris, and D. Rotblatt, *Using the Ground Force Casualty Forecasting System (FORECAS) to Project Casualty Sustainment*, Naval Health Research Center, Report No. 97-39, 1997.

Boeing Defense, Space & Security, "C-17 Globemaster III," Boeing backgrounder, May 2008. As of September 23, 2010:
http://www.boeing.com/defense-space/military/c17/docs/c17_overview.pdf

Brennan, R. J., and K. Rimba, "Rapid Health Assessment in Aceh Jaya District, Indonesia, Following the December 26 Tsunami," *Emergency Medicine Australasia*, Vol. 17, No. 4, August 2005, pp. 341–350.

Bulut, M., R. Fedakar, S. Akkose, H. Ozguc, and R. Tokyay, "Medical Experience of a University Hospital in Turkey After the 1999 Marmara Earthquake," *Emergency Medicine Journal*, Vol. 22, 2005, pp. 494–498.

Defense Health Information Management System, "Medical Analysis Tool (MAT) & Joint Medical Analysis Tool (JMAT)," factsheet, undated. As of August 29, 2011:
http://dhims.health.mil/docs/factsheet-Joint_Medical_Analysis_Tool.pdf

―――, "Joint Medical Analysis Tool (JMAT)," web page, last updated August 22, 2011. As of August 29, 2011:
http://dhims.health.mil/products/theater/jmat.aspx

Doocy, S., C. Robinson, C. Moodie, and G. Burnham, "Tsunami-Related Injury in Aceh Province, Indonesia," *Global Public Health*, Vol. 4, No. 2, March 2009, pp. 205–214.

Fan, S. W., "Clinical Cases Seen in Tsunami Hit Banda Aceh—From a Primary Health Care Perspective," *Annals, Academy of Medicine*, Singapore, Vol. 35, 2006, pp. 54–59.

Greenough, P. G., M. D. Lappi, E. B. Hsu, S. Fink, Y. H. His, et al., "Burden of Disease and Health Status Among Hurricane Katrina–Displaced Persons in Shelters: A Population-Based Cluster Sample," *Annals of Emergency Medicine*, Vol. 51, No. 4, 2008, pp. 426–432.

Haley, T. F., and R. A. De Lorenzo, "Military Medical Assistance Following Natural Disasters: Refining the Rapid Response," *Prehospital and Disaster Medicine*, Vol. 24, No. 1, 2009, pp. 9–10.

Hurd, W. W., and J. G. Jernigan, *Aeromedical Evacuation: Management of Acute and Stabilized Patients*, New York: Springer-Verlag, 2003.

International S.O.S., homepage, undated. As of September 24, 2010:
http://www.internationalsos.com

Kuhn, G. W. S., *CJCS Guide to Battle Casualty Rate Patterns for Conventional Ground Forces*, The Joint Staff, January 15, 1998.

Kwak, Y. H., S. Do Shin, K. S. Kim, W. Y. Kwon, and G. J. Suh, "Experience of a Korean Disaster Medical Assistance Team in Sri Lanka After the South Asia Tsunami," *Journal of Korean Medical Science*, Vol. 21, No. 1, 2006, pp. 143–150.

Liang, N. J., Y. T. Shih, F. Y. Shih, H. M. Wu, H. J. Wang, et al., "Disaster Epidemiology and Medical Response in the Chi-Chi Earthquake in Taiwan," *Annals of Emergency Medicine*, Vol. 38, 2001, pp. 549–555.

Malish, R., D. E. Oliver, R. M. Rush, E. Zarzabal, M. J. Sigmon, et al., "Potential Roles of Military-Specific Response to Natural Disasters—Analysis of the Rapid Deployment of a Mobile Surgical Team to the 2007 Peruvian Earthquake," *Prehospital and Disaster Medicine*, Vol. 24, No. 1, 2009, pp. 3–8.

McNeill, K. M., P. Byers, T. Kittle, S. Hand, J. Parham, et al., "Surveillance for Illness and Injury After Hurricane Katrina—Three Counties, Mississippi, September 5–October 11, 2005," *MMWR Weekly*, Vol. 55, No. 9, March 10, 2006, pp. 231–234.

Mulvey, J. M., S. U. Awan, A. A. Qadri, and M. A. Maqsood, "Profile of Injuries Arising from the 2005 Kashmir Earthquake: The First 72 h.," *International Journal of the Care of the Injured*, Vol. 39, 2008, pp. 554–560.

Owens, B., J. F. Kragh, J. C. Wenke, J. Macaitis, C. Wade, and J. B. Holcomb, "Combat Wounds in Operation Iraqi Freedom and Operation Enduring Freedom," *The Journal of Trauma-Injury, Infection, and Critical Care*, Vol. 64, No. 2, February 2008, pp. 295–299.

Prasartritha, T., R. Tungsiripat, and P. Warachit, "The Revisit of 2004 Tsunami in Thailand: Characteristics of Wounds," *International Wound Journal*, Vol. 5, 2008, pp. 8–19.

Redwood-Campbell, L. J., and L. Riddez, "Post-Tsunami Medical Care: Health Problems Encountered in the International Committee of the Red Cross Hospital in Banda Aceh, Indonesia," *Prehospital and Disaster Medicine*, Vol. 21, No. 1, 2006, pp. 1–7.

Riccardo, F., L. E. Pacifici, A. G. De Rosa, E. Scaroni, L. Nardi, et al., *Epidemiological Surveillance: A Growing Role in Humanitarian Emergencies*, 5th International Congress on Tropical Medicine and International Health, Amsterdam, May 24–28, 2007.

Schultz, J. M., J. Russell, and Z. Espinei, "Epidemiology of Tropical Cyclones: The Dynamics of Disaster, Disease, and Development," *Epidemiologic Reviews*, Vol. 27, 2005, pp. 21–35.

Snyder, D., E. W. Chan, J. J. Burks, M. A. Amouzegar, and A. C. Resnick, *How Should Air Force Expeditionary Medical Capabilities Be Expressed?* Santa Monica, Calif.: RAND Corporation, MG-785-AF, 2009. As of August 29, 2011:
http://www.rand.org/pubs/monographs/MG785.html

Soesman, R., "Humanitarian Assistance Rapid Response Team Package (HARRT) CONOPS," presentation, 13th Air Force, March 10, 2009.

Spinella, P. C., M. A. Borgman, and K. S. Azarow, "Pediatric Trauma in an Austere Combat Environment," *Critical Care Medicine*, Vol. 36, Supplement 7, July 2008, pp. S293–S296.

Streuli, R. A., "Tsunami in South-East Asia—Rapid Response Deployment in Banda Aceh," *Ther Umsch*, Vol. 65, No. 1, January 2008, pp. 15–21.

Teledyne Brown Engineering, Inc., "Tactical Medical Logistics Planning Tool," user web portal, undated. As of September 24, 2010:
http://www.tmlsim.com/portal_4_8/TMLInformation/tabid/53/Default.aspx

Theater Medical Information Program–Joint, "TMIP-J: A Portable Medical Information System," undated. As of August 29, 2011:
http://www.tricare.mil/peo/tmip/TMIP4pgglossy.pdf

TRICARE, homepage, undated. As of September 24, 2010:
http://www.tricare.mil/

U.S. Air Force, Air Force Doctrine Document 2-4.2, *Health Services*, December 2002.

———, Air Force Tactics, Techniques, and Procedures 3-42.5, *Aeromedical Evacuation*, November 2003.

———, Air Force Pamphlet 10-1403, *Air Mobility Planning Factors*, December 18, 2003.

———, Air Force Instruction 11-2AE, *Aeromedical Evacuation Operations Procedures*, May 18, 2005a.

———, Air Force Instruction 11-2AE, *Aeromedical Evacuation Operations Procedures*, Vol. 3, Addenda A: *Aeromedical Evacuation Operations Configuration/Mission Planning*, May 27, 2005b.

U.S. Air Force School of Aerospace Medicine, *EMEDS Reference Handbook*, undated.

U.S. Government Accountability Office, *Homeland Defense: Planning, Resourcing, and Training Issues Challenge DOD's Response to Domestic Chemical, Biological, Radiological, Nuclear, and High-Yield Explosive Incidents*, Washington, D.C., GAO-10-123, October 2009.

U.S. Joint Chiefs of Staff, Joint Publication 4-02, *Health Service Support*, October 2006.

Walker, G. J., J. Zouris, M. F. Galarneau, and J. Dye, *Descriptive Summary of Patients Seen at the Surgical Companies During Operation Iraqi Freedom-1*, Naval Health Research Center, Report No. 04-39, 2004.

Zhang, Lei, He Li, Janis R. Carlton, and Robert Ursano, "The Injury Profile After the 2008 Earthquakes in China," *Injury*, Vol. 40, 2009, pp. 84–86.

Zouris, J., A. Wade, and C. Magno, "Injury and Illness Casualty Distributions During Operation Iraqi Freedom," presentation, Naval Health Research Center, June 2007.

Zouris, J., and G. J. Walker, *Scenario-Based Projections of Wounded-in-Action Patient Condition Code Distributions*, Naval Health Research Center, Technical Report No. 05-32, 2005.